# The Nonlinear Dance: Optimizing Energy Conversion with Piezoelectric Systems

Albert

**Table of Contents**

# Chapter 1. Introduction

This research investigates the complex dynamics of the integrated systems that leverage nonlinearities in both piezoelectromechanical systems and circuits when subjected to complex ambient excitation. Throughout the research, analytical, experimental, and numerical studies are carried out to explore and uncover the complex dynamics of the integrated systems and illuminate optimal strategies for converting DC power.

This chapter provides a survey of previous research and an introduction to the objectives of this research. Five individual chapters will then present the research efforts undertaken. These chapters present the investigations on the integrated nonlinear energy harvesting systems that result in knowledge towards dynamical characteristics and optimal working conditions under real ambient environments. Then the final chapter concludes the major discoveries of the research and discusses potential applications and future work directions.

## 1.1 Literature review

Among all kinds of ambient energy sources, vibration energy is a desirable choice to harvest for the abundance and persistence of kinetic energy in applications. Piezoelectric beams are widely utilized to transfer this kinetic energy to electrical power due to the high-power density of piezoelectric material [1]. A piezoelectric structure with linear dynamics is first utilized to harvest vibration energy by tuning its resonance frequency to convert maximum power from the kinetic energy sources [1] [2] [3]. Because of the unfavorable narrow-band characteristic of linear energy harvesters, a wide variety of methods, such as energy harvester array, mechanical stoppers, and nonlinearity, have been proposed to broaden the effective bandwidth [4] [5] [6] [7] [8].

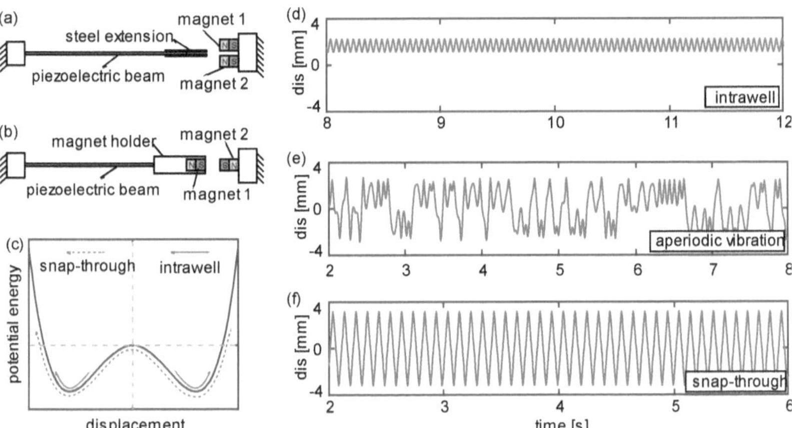


Figure 1.1 Schematics of piezomagnetoelastic systems with (a) attractive magnetic force and (b) repulsive magnetic force. (c) potential energy profile of bistable oscillators. Time series of (d) intrawell vibration, (e) aperiodic vibration, and (f) snap-through vibration.

Among them, nonlinear systems, especially bistable systems, are preferable for the non-resonant nature as the large amplitude vibration can be triggered even by relatively low input vibrations in a wide frequency band [9] [10] [11] [12] [13]. Figure 1.1(a) and (b) respectively presents two approaches to construct piezomagnetoelastic systems with bistablity. In Figure 1.1(a), a pair of neodymium magnets is positioned near the ferromagnetic extension to introduce an attractive magnetic force [14]. In comparison, two

2

magnets are installed opposingly to include a repulsive magnetic force for the piezomagnetoelastic system in Figure 1.1(b) [15]. By adjusting the magnet positions, the piezomagnetoelastic system may demonstrate a unique double-well restoring force potential as depicted in Figure 1.1(c) [8]. With two stable equilibrium positions, three distinct dynamic regimes may be determined according to the input amplitude [8]. For comparative small excitation, the inertial mass may oscillate around one stable equilibrium to generate small-amplitude intrawell vibration characterized in Figure 1.1(d). Alternatively, the increased input may generate an aperiodic oscillation in Figure 1.1(e). The third dynamic regime named snap-through vibration is demonstrated in Figure 1.1(f). Among the three dynamic regimes, snap-through vibrations that oscillate through two stable equilibria as indicated in Figure 1.1(c) are favorable for energy harvesting by virtue of the large amplitude vibrations.

One assumption often adopted while studying the superiority of nonlinear energy harvesters is symmetry in potential energy functions [12] [16] [15] [17]. Yet, due to structure imperfections, asymmetric applied magnetic field, or the existence of static load, asymmetry is unavoidable in the potential energy profiles of vibration energy harvesters [18] [19]. On the other hand, pure harmonic or pure stochastic excitation is usually employed in studies to characterize electrodynamic responses [15] [19] [20]. In fact, the environments that harvesters are deployed may experience complex vibration conditions rather than pure harmonic or stochastic excitation, such as combined harmonic and stochastic excitation [21] [22] or impulsive excitation [23] [24] [25]. In addition, since microelectronic devices require a regulated direct current (DC) input voltage, a rectifying circuit is necessary for energy harvesters to convert the oscillating alternating current (AC) signals to DC power. Yet, due to the complicated mechanical-electrical coupling in energy harvesters, analytical approaches seldom address DC power delivery for the nonlinear energy harvester. Consequently, there is a fundamental void in understanding the non-smooth dynamics of nonlinear energy harvester interfacing with a rectification circuit in complex vibration environments for maximizing the harvested DC power.

On the other hand, numerous researchers concentrated on the nonlinearity and non-smoothness of harvesting circuits to improve power captured from the linear vibration energy harvester. By utilizing switching strategies with the standard rectification, such as the parallel-SSHI (synchronized switching harvesting on inductor) circuit and SCE (Synchronized Charge Extraction) circuit, the harvesting power can be substantially enhanced. In order to achieve optimal power for a selected circuit, rigorous impedance analysis for a variety of rectification circuits is conducted to identify the impedance matching condition for a linear energy harvester [26] [27] [28]. Despite the enhancements that have been achieved from the circuits' design, these advancements are rarely reported with vibration energy harvesters operated in a nonlinear regime.

Moreover, the ambient kinetic energy is unpredictable, although abundant [29]. As such, the power delivered from the harvester may be insufficient occasionally for devices. An energy storage component is essential to separate the process of harvesting energy and powering devices [30] [31]. Supercapacitor and rechargeable battery are two common energy reservoirs. Although capacitors are generally employed in the harvesting circuits, most of them function as smoothing units to reduce the ripples of DC voltage across the resistance. Therefore, the knowledge in understanding the influences of energy storage components on power delivery is still lacking.

Considering the state-of-the-art advancements in vibration energy harvesting communities, the insights on optimal strategies are still limited since nonlinear features are only considered in either vibration structures or electrical platforms. As a result, rigorous efforts should be devoted to connecting the knowledge on system-level implementation to understand the optimal combination of the multiple structural, electrical, and material physics under complex ambient environments. Moreover, energy harvesting systems need to deliver the required DC power to electrical load over a long-time horizon. Most design practices for energy harvesters often report strategies based on maximizing output voltage or wide frequency range [32] [33] [34] [35]. Yet, such enhanced dynamic behavior may correspondingly be detrimental to the integrity of harvesters. Therefore, more attention should be given to the performance-balanced functioning of harvesters to deliver excellent DC power in a broad frequency range with exceptional mechanical integrity.

## 1.2 Objectives of the book

This research investigates the intricate dynamics of integrated systems leveraging mechanical and electrical nonlinearities in complex vibration environments. The revealed principles bridge the advancements between the research communities in vibration energy harvesting and create understandings of techniques to capitalize on synergistic dynamic behavior.

To achieve the goal, four objectives are addressed here.

- Characterization of the dynamics of integrated energy harvesting systems in complex vibration environments

The dynamics of integrated energy harvesting systems under complex vibration environments are investigated through analytical and experimental efforts. The revealed principles create new knowledge in understanding the complex dynamics and assist in the real-world implementation of energy harvesters for improved, robust power delivery.

- Bridge the advancements between the research communities in vibration energy harvesting

Analytical approaches have been established to accommodate the mechanical and electrical nonlinearities to facilitate insights for the first time on the suitable integration of these advanced sub-systems. Beyond vibration energy harvesting, the theoretical tools may be leveraged in future studies where multiphysics combine towards the resulting system dynamics, and as such is valuable for applied mathematicians.

- Demonstration of the influences of energy storage components on power delivery from integrated energy harvesting systems

Analytical and numerical approaches have been employed to investigate and compare the influences of energy storage units on power delivery from various integrated energy harvesting systems. These findings facilitate insights on optimal integration of the multiphysics sub-systems to assist in developing sustainable IoT devices.

- Establishment of an optimization technique for future performance-balanced energy harvesting system design

An optimization technique is established to exploit nonlinearity in the design of custom-shaped harvesters to deliver excellent DC output power in a broad frequency range with exceptional mechanical integrity. The coupled influences of nonlinearity and laminated beam shape are illuminated for the first time to assist in the deployment of performance-balanced functioning systems.

## 1.3 Organization of the book

The chapters of the book are organized as follows.

Chapter 2 examines the dynamics of integrated nonlinear energy harvesting systems under combined harmonic and stochastic excitation analytically and experimentally. The influences of asymmetry and excitation amplitude on the electrodynamic responses and DC power delivery have been demonstrated to help develop the deployment techniques for nonlinear vibration energy harvesters in practical excitation environments.

Chapter 3 employs physics-based and data-driven analyses to discover the underlying relationships between the impulse-induced nonlinear dynamics and the result converted electrical energy. The findings demonstrate a successful synthesis of data-driven and physical models to guide attention to optimal designs of nonlinear energy harvesting systems subjected to periodic impulse excitation.

Chapter 4 presents the investigations on impedance to examine the optimal DC power operation of a nonlinear energy harvester interfaced with a buck-boost converter for practical electrical power management. These results suggest the implementation of optimal nonlinear energy harvesting systems may find effective guidance through power flow concepts.

Chapter 5 investigates the effects of energy storage components on power delivery and compares the performance of switch-controlled harvesting circuits when integrating with nonlinear energy harvesters. These findings aim to demonstrate optimal strategies when the energy storage component is incorporated.

Chapter 6 introduces a multi-objective optimization tool to exploit nonlinearity in the design of custom-shaped harvesters in terms of output power, broad bandwidth, and mechanical integrity. Experimental and analytical efforts have been carried out to validate the design practices uncovered from optimization.

Chapter 7 concludes this book with a summary of the key discoveries and discussion of potential applications.

**Chapter 2.  Characterization of challenges in asymmetric nonlinear vibration energy harvesters subjected to realistic excitation**

With an explosion of the internet of things, vibration energy harvesting provides an environmentally friendly solution to replace consumable batteries in powering IoT wireless sensors. Yet, when implemented in practice, ambient vibration input energy is much less periodic than assumptions adopted in previous studies. This becomes especially important given that asymmetries are inevitable in nonlinear device platforms. This research sheds light on these complex challenges of practical vibration energy harvesting by developing and exploring an analytical model based on equivalent linearization. The modeling approach provides an opportunity to understand influences of asymmetry, nonlinearity, and combined excitation response on the DC power delivery of energy harvesters. In the analytical model, a weighted Gaussian joint distribution is utilized to approximate the influences caused by the random excitation. Combined with numerical and experimental validation, the analysis indicates that with the increase of stochastic base acceleration, two outcomes are possible. A first outcome involves an enhancement of DC power by way of triggering large amplitude nonlinear oscillations. A second outcome corresponds to a loss of high power delivery since the noise interferes with the attainment of the snap-through dynamic. Either reducing asymmetry or increasing harmonic excitation component is found to be favorable to induce the power-enhancing dynamics and inhibit the occurrence of the second case. Although with the simplified Gaussian distribution, the analytical framework cannot reproduce exact details of the dynamic responses in every case, the results show that the statistical trends of the analysis are overall borne out in simulation and experiment. This indicates the new modeling of this research may help guide attention to design and deployment techniques for nonlinear vibration energy harvesters in practical combined excitation environments, where limitations on precise manufacture or placement may introduce structural asymmetry.

## 2.1 Introduction

The extensive network of interconnected wireless devices termed the Internet-of-Things has been widely built up in recent years for applications such as building operations [36], healthcare [37], and smart farming [38], to name a few. The IoT results in smart, data-informed strategies by continuously collecting and processing trends of user consumption, system health, and other important metrics. With a projected increase of IoT devices to a trillion by 2025 [39], the high demand on disposable batteries suggests a clear threat to the sustainability of our environment and infrastructure. With the decrease of required power for IoT devices to microwatts [39], vibration energy harvesting can be an environmentally friendly energy provider of on-demand electrical power due to the high power density and broad availability of kinetic energy [40].

In recent years, numerous research efforts have been devoted to improving the performance of vibration energy harvesting systems. Because of the narrowband characteristic of linear vibration energy harvesters, nonlinear energy harvesting systems are proposed to meet the requirements of broadband frequency sensitivity and high output power resulting from the input vibration energy [9] [41] [15] [42]. Because of the multistability feature of energy harvesters with particular nonlinearities, a wide variety of methods have been proposed to readily attain high energy orbit vibration [43] [44] [16]. For example, Zhou et al. [16] proposed a flexible bistable energy harvester with a controllable potential energy function that helps one govern the potential energy barrier for triggering snap-through vibration. Wang and Liao [44] utilized the load perturbation method to create strategies that transform system states from intrawell vibration to snap-through oscillation.

In addition, optimization strategies are applied seeking energy harvesting system designs with maximized performance [34] [45] [46]. Dietl and Garcia [32] employed a gradient search optimization tool to determine that a curved beam shape along the beam length axis provides optimal and uniform strain distribution for power maximization. Cai and Harne [47] [48] utilized the genetic algorithm optimization method to uncover the entangled influences of nonlinearity, beam shape, and tip mass for high performing and structurally resilient energy harvesting cantilever designs. Since rectification is necessary to convert the AC voltage from the piezoelectric beam to DC voltage, studies on harvesting circuits explore methods for maximizing power delivery [49] [50]. By utilizing switching strategies with standard rectification, the synchronous electric charge extraction (SECE) circuit was found to yield a 400% increase in harvested power [51]. On the basis of the SECE circuit, Lefeuvre et al. [52] presented a phase-shift SECE (PS-SECE) circuit to improve the delivered power up to a theoretical limit for piezoelectric devices having high electromechanical coupling.

From this survey of state-of-the-art investigations, one common assumption adopted is the symmetry of potential energy functions when leveraging nonlinearities for energy harvesting [41] [15] [47]. Yet, due to factors such as asymmetric magnetic fields, structure imperfections, or static and gravitational loads, asymmetry is nearly unavoidable in implementing nonlinear energy harvesting systems. This indicates it is necessary to investigate the influences of asymmetry in the development of nonlinear energy harvesters. He and Daqaq [19] examined the AC output power of asymmetric nonlinear energy harvesting systems subjected to pure white noise excitation vibration and indicated that the existence of asymmetry may increase the output power for a monostable system, whereas the performance of a bistable system is deteriorated. Wang et al. [18] introduced an installation bias angle to asymmetric bistable energy harvesting systems to reduce negative effects caused by asymmetric potentials for the energy harvesters when subjected to pure harmonic or random excitation.

Moreover, pure harmonic or pure stochastic excitation is usually employed in studies to characterize electrodynamic responses [20] [12]. In fact, environments wherein harvesters may be employed to sustain IoT devices may experience complex vibration conditions such as combined harmonic and stochastic excitation [21] [22]. For instance, with the influence of moving vehicles, the vibrations of bridges show a combined influence of periodic oscillation and random vibration [22]. In order to analytically predict the dynamical characteristics when subjected to combined harmonic and stochastic excitation, methods such as Gaussian closure method [53] [54], stochastic averaging method [55] [56], and equivalent linearization method [57], have been applied to symmetric Duffing oscillators to statistically characterize the structural dynamics. Yet, due to the complicated mechanical-electrical coupling in energy harvesters, analytical approaches seldom address DC power delivery that results following rectification stages. A few insights on this class of structural-electrical coupling have been obtained. For instance, Kim et al. [58] used numerical approaches to identify the stochastic resonance phenomenon for rotating energy harvesting systems under combined harmonic and stochastic excitation and validated the findings by experiments. Dai and Harne [14]introduced an analytical approach to determine the relationship between the electrical and mechanical responses and thus predict the electromechanical responses of cantilevered energy harvesters due to combined harmonic and stochastic excitation. On the other hand, no works have synthesized a prediction tool for the intricate electromechanical responses of asymmetric nonlinear vibration energy harvesters subjected to combined vibration excitation conditions. The insight from such an analytical process would be strongly relevant to help transition energy harvesting devices to practice.

Motivated to fill the void in understanding, this research investigates the dynamic behaviors of an asymmetric nonlinear energy harvester integrated with a standard rectification harvesting (SEH) circuit. The aim is to characterize the mechanical and electrical responses when subjected to combined harmonic and stochastic excitation. In the following sections, an analytical framework is first established for the proposed system. After validating the system numerically and experimentally, the influences of asymmetry are investigated considering the combined harmonic and stochastic excitation condition. Finally, a summary of main findings is provided to conclude this chapter.

## 2.2 Analytical model formulation

The vibration energy harvesting platform considered is shown in Figure 2.1. A piezoelectric cantilever has a tip mass $M_0$ constructed by the magnet 1 and its holder at the free end. The beam is excited by base motion representative of the ambient kinetic energy. The electrodes of the beam are interfaced with a SEH circuit to convert the AC voltage to a DC voltage. A pair of repulsive magnets is utilized to introduce nonlinearities into the system. The gap $d_2$ and bias $\Delta$ between the two magnets determine the magnetic force $F_m$ that governs the type of nonlinearity and the significance of asymmetry induced into the potential energy profile of the system. The polynomial expression in Eq. (2.1) is adopted in the model to approximate the nonlinear magnetic force $F_m$ [59] [60]. The parameters $k_1, k_2$ and $k_3$ are experimentally identified, as described in Section 2.3.

$$F_m = k_1 \underline{x} + k_2 \underline{x}^2 + k_3 \underline{x}^3 \tag{2.1}$$

Therefore, the non-dimensional equivalent lumped parameter governing equations for the lowest order structural and electrical responses are [19] [14]:

$$x'' + \eta x' + (1-p)x + \beta_2 x^2 + \beta_3 x^3 + \kappa v_p = -x_a'' \tag{2.2a}$$

$$v_p' + i_p = \theta x' \tag{2.2b}$$

The relationships between the non-dimensional parameters with the physical system parameters are

$$x = \underline{x}/x_0 \ ; \ v_p = \underline{v}_p/V_0 \ ; \ \omega_0 = \sqrt{k/m} \ ; \ \tau = \omega_0 t \ ; \ p = -k_1/k \ ; \ \beta_2 = k_2 x_0^2/k \ ; \ \beta_3 = k_3 x_0^2/k \ ; \ \kappa = \alpha V_0/(k x_0) \ ;$$

$$\eta = d/m\omega_0 \ ; \ \theta = \alpha x_0/(C_p V_0) \ ; \ i_p = \underline{i}_p \omega_0/(C_p V_0) \ ; \ x_a'' = \underline{\ddot{x}}_a x_0 k/m \ ; \tag{2.3a-l}$$

Here, $x$ is the beam tip displacement relative to the motion of the base displacement $x_a$; $m$, $d$, and $k$ are the equivalent lumped mass, viscous damping, and linear stiffness corresponding to the first mode of the beam vibration; $p$ is the load parameter indicating the influence of magnetic forces on reducing the linear stiffness $k$; $\alpha$ is the electromechanical coupling constant; $C_p$ is the internal capacitance of the piezoelectric layers in the cantilever; $v_p$ is the voltage across the piezoelectric beam electrodes; $i_p$ is the corresponding current through the harvesting circuit; $x_0$ and $V_0$ are characteristic length and voltage, which are 1 mm and 1 V respectively; The over dot and apostrophe operators indicate differentiation with respect to time $t$ and non-dimensional time $\tau$, respectively.

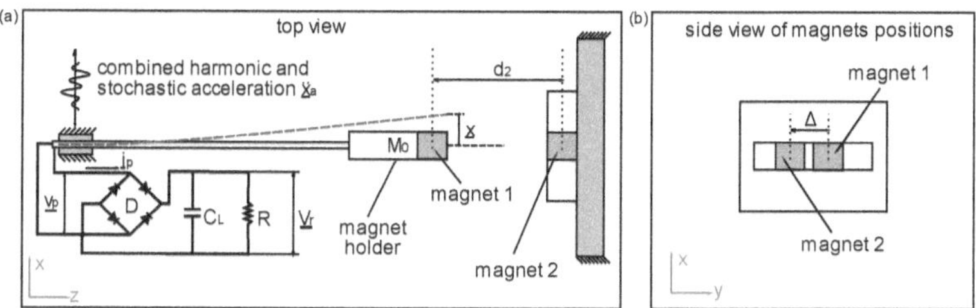


Figure 2.1 Schematic of nonlinear energy harvesting system

The non-dimensional combined harmonic and stochastic base acceleration is

$$-x_a'' = a\cos\omega\tau + \sigma w(\tau) \tag{2.4}$$

Here, the non-dimensional standard deviation of the stochastic acceleration is $\sigma$, while the amplitude of harmonic base excitation is $a$; $\omega$ is the non-dimensional angular frequency of the harmonic base excitation determined by $\omega = \underline{\omega}/\omega_0$; $\underline{\omega}$ is the absolute angular frequency; $w(\tau)$ is a unit Gaussian white noise process.

For the nonlinear structural system expressed by Eq. (2.2), the mean value of the displacement response $x$ is defined to be $m_x$. Consequently, the displacement response $x$ is represented by

$$x = x_0 + m_x \tag{2.5}$$

Here, $x_0$ is the zero mean dynamic response.

The equivalent linearization method [19] [57] [61] is applied to Eq. (2.2a) to linearize the quadratic and cubic nonlinearities by introducing the equivalent linear natural frequency $\omega_e$ and displacement offset $\varepsilon$.

$$x_0'' + \eta x_0' + \omega_e^2 x_0 + \varepsilon + \kappa v_p = -z'' \tag{2.6a}$$

$$v_p' + i_p = \theta x_0' \tag{2.6b}$$

In order to ensure the equivalence between Eq. (2.2) and Eq. (2.6), the mean-square error between (2.2) and (2.6) is minimized using (2.7a) and (2.7b) [19] [57] [14] [61].

$$\frac{\partial \langle E^2 \rangle}{\partial \omega_e^2} = \frac{\partial}{\partial \omega_e^2} \left\langle \left[ (1-p)x + \beta_2 x^2 + \beta_3 x^3 - \omega_e^2 x_0 - \epsilon \right]^2 \right\rangle = 0 \tag{2.7a}$$

$$\frac{\partial \langle E^2 \rangle}{\partial \varepsilon} = \frac{\partial}{\partial \varepsilon} \left\langle \left[ (1-p)x + \beta_2 x^2 + \beta_3 x^3 - \omega_e^2 x_0 - \epsilon \right]^2 \right\rangle = 0 \tag{2.7b}$$

The bracket $\langle \rangle$ indicates the mathematical expectation or time-averaged value. By calculating the mean value for both sides of Eq. (2.6a), it is found that $\varepsilon = 0$.

Based on the linearized system Eq. (2.6) with equivalent frequency $\omega_e$ and offset $\varepsilon$ determined through the Eq. (2.7), superposition is applied. Here, the total structural and electrical responses are approximated to be the summation of responses individually attributed to the harmonic or stochastic excitation components.

$$x_0 = x_h + x_r \tag{2.8a}$$

$$v_p = v_{ph} + v_{pr} \tag{2.8b}$$

The $x_h$ and $v_{ph}$ represent the structural and electrical responses, respectively, of the equivalent linear system resulting from the harmonic excitation component, as governed by Eq. (2.9). The $x_r$ and $v_{pr}$ are the stochastic components of the displacement and voltage specifically governed by Eq. (2.10).

$$x_h'' + \eta x_h' + \omega_e^2 x_h + \kappa v_{ph} = a \cos \omega\tau \tag{2.9a}$$

$$v_{ph}' + i_{ph} = \theta x_h' \tag{2.9b}$$

$$x_r'' + \eta x_r' + \omega_e^2 x_r + \kappa v_{pr} = \sigma w(\tau) \tag{2.10a}$$

$$v_{pr}' + i_{pr} = \theta x_r' \tag{2.10b}$$

For the structural response $x_h$ due to the harmonic excitation, the fundamental frequency $\omega$ of the base acceleration is assumed to be the dominant frequency of displacement. Thus the $x_h$ is given by

$$x_h = h \sin \omega\tau + g \cos \omega\tau = n \cos(\omega\tau - \varphi) \tag{2.11}$$

The corresponding piezoelectric voltage $v_{ph}$ is represented by a piecewise function shown in Eq. (2.12) [14] [26].

$$v_{ph} = \begin{cases} \theta n (\cos \omega\tau - 1) + V_{rh}; & 0 < \omega\tau \le \Theta \\ -V_{rh}; & \Theta < \omega\tau \le \pi \\ \theta n (\cos \omega\tau + 1) - V_{rh}; & \pi < \omega\tau \le \pi + \Theta \\ V_{rh}; & \pi + \Theta < \omega\tau \le 2\pi \end{cases} \tag{2.12}$$

The $V_{rh}$ is the non-dimensional magnitude of the rectified voltage resulting from the load resistance $R$, shown in Figure 2.1.

The magnitude of the rectified voltage is obtained by integrating Eq. (2.6b) over a semi-period of the harmonic excitation [14] [62],

$$V_{rh} = \frac{2\theta n}{\left(\pi/\omega\right)\rho + 2} \ , \ \rho = 1/C_p R \omega_0 \tag{2.13a-b}$$

Here, $\rho$ is the non-dimensional resistance.

Consequently, the piezoelectric voltage $v_{ph}$ is related to the lowest order displacement $x_h$ and velocity $x_h'$ by the fundamental term of a Fourier series of Eq. (2.12).

$$v_{ph} = \frac{A}{\omega} x_h' + B x_h \tag{2.14}$$

where

$$A = \frac{1}{\pi}\theta\sin^2\Theta \ , \ B = \frac{1}{2\pi}\theta\left(2\Theta - \sin 2\Theta\right), \text{ and } \cos\Theta = \frac{\pi - 2\omega\omega_0 C_p R}{\pi + 2\omega\omega_0 C_p R} . \tag{2.15a-c}$$

The response $x_h$ is achieved by substituting Eqs. (2.11) and (2.14) into Eq. (2.9a).

$$g = \frac{a\left(B\kappa - \omega^2 + \omega_e^2\right)}{\left(B\kappa - \omega^2 + \omega_e^2\right)^2 + \left(A\kappa + \eta\omega\right)^2} ; \ h = \frac{a\left(A\kappa + \eta\omega\right)}{\left(B\kappa - \omega^2 + \omega_e^2\right)^2 + \left(A\kappa + \eta\omega\right)^2} ; \tag{2.16a-b}$$

The corresponding electrical responses are also determined by Eqs. (2.13) and (2.14) following computation of Eq. (2.16).

For the stochastic responses governed by Eq. (2.10), based on the generalized harmonic function [63], the corresponding stochastic displacement $x_r$ and velocity $x_r'$ are written as

$$x_r = n_r \cos\left(\omega_r \tau + \varphi_r\right) = h_r \sin\omega_r\tau + g_r \cos\omega_r\tau \tag{2.17a}$$

$$x_r' = -n_r\omega_r \sin\left(\omega_r\tau - \varphi_r\right) = h_r\omega_r \cos\omega_r\tau - g_r\omega_r \sin\omega_r\tau \tag{2.17b}$$

18

Here $\omega_r$ and $n_r$ are the angular frequency and amplitude for the periodic non-stationary process [20] [63].

Consequently, the relationship shown in Eq. (2.14) is also applicable for stochastic components according to the frequency $\omega_r$.

$$v_{pr} = \frac{A_r}{\omega_r} x_r' + B_r x_r \tag{2.18}$$

where

$$A_r = \frac{1}{\pi}\theta \sin^2\Theta_r, \ B_r = \frac{1}{2\pi}\theta(2\Theta_r - \sin 2\Theta_r), \text{ and } \cos\Theta_r = \frac{\pi - 2\omega_r\omega_0 C_p R}{\pi + 2\omega_r\omega_0 C_p R} \tag{2.19a-c}$$

Substituting Eq. (2.19) into Eq. (2.10a), the equivalent mechanical governing equation under the stochastic excitation is obtained.

$$x_r'' + \left(\eta + \frac{A_r\kappa}{\omega_{nr}}\right)x_r' + \left(\omega_e^2 + \kappa B_r\right)x_r = \sigma w(\tau) \tag{2.20}$$

From Eq. (2.20), the non-stationary angular frequency is [64]

$$\omega_r = \sqrt{\left(\omega_e^2 + \kappa B_r\right)} \tag{2.21}$$

To determine the mean-square of the stochastic displacement component $\langle x_r^2 \rangle$ from Eq. (2.20), the probability density distribution must first be evaluated. Therefore, assuming a standard Gaussian distribution for $x_r$, the probability distribution for the random process $x_r$ considering a mean displacement $m_x$ is given by Eq. (2.22).

$$f(x_{rm}) = g(x_{rm} | m_x, \sigma_{xg}) = \frac{1}{\sigma_{xg}\sqrt{2\pi}} e^{-\frac{1}{2}\left(\frac{x_{rm} - m_x}{\sigma_{xg}}\right)^2} \tag{2.22a}$$

$$\sigma_{xg} = \frac{\sigma}{\sqrt{2\left(\eta + \dfrac{A_r \kappa}{\omega_{nr}}\right)\left(\omega_e^2 + \kappa B_r\right)}}$$ (2.22b)

Here, $x_{rm}$ is the summation of $x_r$ and $m_x$, $\sigma_{xg}$ is the non-dimensional standard deviation of $x_{rm}$ associated with the standard Gaussian distribution and determined by Eq. (2.22b). Yet, for nonlinear systems with multiple static equilibria, a standard Gaussian distribution assumption for the random variable $x_r$ does not provide good approximation of observable response statistics [65]. Therefore, to improve the accuracy in the analytical predictions, a weight $w_e$ is introduced to formulate a weighted Gaussian distribution in Eq. (2.23).

$$f_1(x_{rm}) = g\left(x_{rm}\middle|m_x,\sigma_{xw}\right) = g\left(x_{rm}\middle|m_x,w_e\sigma_{xg}\right)$$ (2.23)

The $\sigma_{xw}$ corresponds to the standard deviation of $x_{rm}$ associated with the weighted Gaussian distribution. In other words, $\sigma_{xw} = w_e\sigma_{xg}$, where the weight $w_e$ is a product with the non-dimensional standard deviation $\sigma_{xg}$ of $x_{rm}$.

In order to determine the weight, a probability density distribution for nonlinear systems under pure stochastic excitation with multiple stable equilibria is given in Eq. (2.24) [65] against which the weighted Gaussians Eq. (2.23) are compared ignoring the influence of harmonic excitation.

$$f_2(x_{rm}) = \frac{1}{2}g\left(x_{rm}\middle|x_1,\sigma_{xg}\right) + \frac{1}{2}g\left(x_{rm}\middle|x_2,\sigma_{xg}\right)$$ (2.24)

Here in Eq. (2.24) the $g\left(x_{rm}\middle|x_1,\sigma_{xg}\right)$ and $g\left(x_{rm}\middle|x_2,\sigma_{xg}\right)$ are standard Gaussian distributions with mean values of $x_1$ and $x_2$, such that $x_1$ and $x_2$ are the statically stable equilibria of the nonlinear system. To simplify the following discussion, the radicand in Eq. (2.22b) is considered to be a unit valued constant towards determining the suitable weight $w_e$.

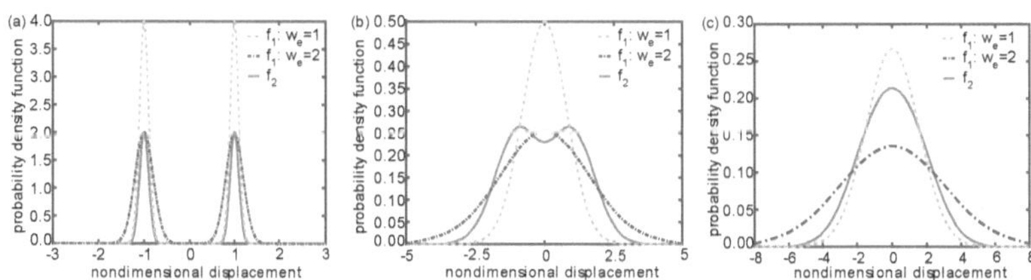

Figure 2.2 Probability density distributions for standard deviations of stochastic base acceleration of (a) $\sigma = 0.1$, (b) $\sigma = 0.8$, and (c) $\sigma = 1.5$.

Figure 2.2 presents the probability density distributions given by Eqs. (2.23) and (2.24). For the probability distribution shown in Eq. (2.24) $f_2$, at noise standard deviation $\sigma = 0.1$ in Figure 2.2(a), the two peak values are thoroughly separated, which indicates that snap-through vibration rarely occurs as caused by the random excitation component. This is because snap-through vibration is indicative of a zero mean value. With the increase of noise standard deviation to 0.8 as shown in Figure 2.2(b), the two peaks begin to coalesce, which indicates that stochastic excitation may trigger snap-through vibration more often. At much higher noise standard deviation, Figure 2.2(c) reveals that the two peaks merge so that there is high probability that the mean of the displacement is zero, corresponding to snap-through vibration.

For the distribution shown in Eq. (2.23) $f_1$, two weights are selected, $w_e = 1$ and $w_e = 2$ for evaluation in Figure 2.2. When the noise intensity is too low to trigger snap-through vibrations, Figure 2.2(a) shows that the mean value $m_x$ is statistically identical to the value of the equilibria. Yet, for increase in the noise standard deviation $\sigma$, such as the values $\sigma = 0.8$ and $\sigma = 1.5$ in Figures 2.2(b,c), snap-through vibration happens more frequently and the mean value $m_x$ takes the mean of the two equilibria, which is approximately zero.

As shown through the results of Figure 2.2, the distribution of Eq. (2.23) $f_1$ may sufficiently reproduce the probability distribution of the exact nonlinear system based on the use of weight $w_e$. Compared with the distributions $f_2$ determined by Eq. (2.24), the weight $w_e = 2$ applied to distribution $f_1$ leads to a sufficient approximation of the peak value of the probability density function $f_2$.

To quantify the agreement of the distributions, Table 2.1 summarizes the mean-square values of the non-dimensional displacement by numerically integrating the expressions Eqs. (2.23) and (2.24). At low noise intensity in Figure 2.2(a), the integral interval is chosen to be [-1.5, 1.5]. In comparison, for noise intensities in Figures 2.2(b,c), the range [-3, 3] is selected. The relative errors of the approximated distributions $f_1$ with respect to the accurate distribution $f_2$ are provided in Table 2.1. It is observed that the distributions $f_1$ adequately approximate the distribution $f_2$ when the noise standard deviation is $\sigma = 0.1$ or $\sigma = 1.5$. Yet, for the intermediate standard deviation of stochastic excitation $\sigma = 0.8$, at which point Figure 2.2 shows that the two peaks of statistical response coalesce, the weighted

Gaussian distribution $f_1$ with the weight $w_e = 2$ better agrees with the accurate distribution $f_2$.

Table 2.1 Mean-square value of non-dimensional displacement at three non-dimensional noise standard deviations.

| Noise standard deviation | $f_1 : w_e = 1$ | | $f_1 : w_e = 2$ | | $f_2$ |
| --- | --- | --- | --- | --- | --- |
| | Mean-square displacement value | Relative error | Mean-square displacement value | Relative error | Mean-square displacement value |
| $\sigma = 0.1$ | 1.0010 | 2.32% | 1.0010 | 2.32% | 1.0248 |
| $\sigma = 0.8$ | 0.6382 | 59.45% | 1.7440 | 10.82% | 1.5737 |
| $\sigma = 1.5$ | 1.6617 | 13.58% | 1.7887 | 7.34% | 1.9229 |

With the probability density distribution given in Eq. (2.24), the mean square value for the variable $x_{rm}$ can be calculated by the Gaussian closure method when subjected to pure stochastic excitation [65]. Yet, when considering the combined harmonic and stochastic excitation, there is no direct way to establish an analytical expression for the mean square value of $x_{rm}$ without an assumption for the Gaussian distribution [66]. Therefore, in the study the weighted Gaussian distribution $f_1$ with $w_e = 2$ in Eq. (2.23) is chosen to balance the accuracy and simplicity in predicting the statistical responses analytically. The analytical expression for the mean values of random variable $x_r$ is then given in Eq. (2.25), where $w_e = 2$.

$$\langle x_r^2 \rangle = \frac{(w_e \sigma)^2}{2\left( \eta + \dfrac{A_r \kappa}{\omega_{nr}} \right)(\omega_e^2 + \kappa B_r)} \tag{2.25}$$

The corresponding mean-square rectified voltage is approximated from Eq. (2.13) to be

$$< v_r^2 > = \frac{4\theta^2 \omega_{nr}^2 \langle x_r^2 \rangle}{\pi^2 \rho^2 + 4\omega_{nr}^2 + 4\pi \rho \omega_{nr}} \tag{2.26}$$

Assuming the relationships in Eq. (2.27), the expressions for $\omega_e^2$ and $\varepsilon$ are determined by substitution and simplification of the Eqs. (2.5), (2.8), and (2.11) into Eq. (2.7). The squared equivalent frequency $\omega_e^2$ and offset $\varepsilon$ are given in Eq. (2.28).

$$\langle x_r \rangle = 0; \ \langle x_r^2 \rangle = \langle x_r^2 \rangle; \ \langle x_r^3 \rangle \approx 0; \ \langle x_r^4 \rangle \approx 3 \langle x_r^2 \rangle^2 - 2 \langle x_r \rangle^4; \tag{2.27}$$

$$\omega_e^2 = \frac{1}{\left(2\langle x_r^2 \rangle + n^2\right)} \left( \begin{array}{l} 2\langle x_r^2 \rangle + n^2 - 2\langle x_r^2 \rangle p - n^2 p + 6\langle x_r^2 \rangle^2 \beta + 6\langle x_r^2 \rangle m_x^2 \beta_3 + 6\langle x_r^2 \rangle n^2 \beta_3 + \\ 3m_x^2 n^2 \beta_3 + \dfrac{3n^4 \beta_3}{4} + 4\langle x_r^2 \rangle m_x \beta_2 + 2m_x n^2 \beta_2 \end{array} \right) \tag{2.28a}$$

$$\varepsilon = m_x - m_x p + m_x^3 \beta_3 + \frac{3}{2} m_x n^2 \beta_3 + 3m_x \langle x_r^2 \rangle \beta_3 + m_x^2 \beta_2 + \frac{1}{2} n^2 \beta_2 + \langle x_r^2 \rangle \beta_2 \tag{2.28b}$$

Studying the expressions of $\omega_e^2$ and $\varepsilon$ in Eq. (2.28), despite the perceived independence of the harmonic structural response $x_h$ and the stochastic structural response $x_r$, mutual influences exist through the coupling between harmonic and stochastic components evident in the equivalent linearized parameters in Eq. (2.28). Consequently, Eqs. (2.16), (2.25) and (2.28) are solved simultaneously. The corresponding electrical responses are then determined based on Eqs. (2.13) and (2.26). Due to the contribution from the stochastic base acceleration, the total mean-square value of converted voltage shown in Eq. (2.29) is taken to characterize the energy harvesting output in the following investigations.

$$\left\langle v_r^2 \right\rangle = v_{rh}^2 + \left\langle v_{rr}^2 \right\rangle \tag{2.29}$$

## 2.3 Experimental systems identification

The experimental platform is shown in Figure 2.3. A piezoelectric beam with parallel-wired PZT layers (PPA-2014; Mide Technology) is clamped to an aluminum base, which is affixed to the electrodynamic shaker table (APS Dynamics 400). At the free end of the beam, magnet 1 is secured in an aluminum magnet holder connected to the beam tip. The nonlinearity and asymmetry included in the system are adjusted by the position of magnet 2 in relation to magnet 1. Two displacement lasers (Micro-Epsilon ILD-1420) are utilized to measure the absolute displacement of the shaker table and beam tip. The electrodynamic shaker table is driven by an amplifier (Crown XLS 2500) fed appropriate combinations of harmonic and stochastic excitation voltage. An accelerometer (PCB Piezotronics 333B40) on the shaker table is used to confirm the frequency content of the resulting base acceleration. A standard rectifier bridge (1N4148 diodes) is connected to the piezoelectric beam to convert the AC to DC signal. Following the bridge, a smoothing capacitor $C_r$ and resistive load $R$ are utilized to quantify the harvested electric energy.

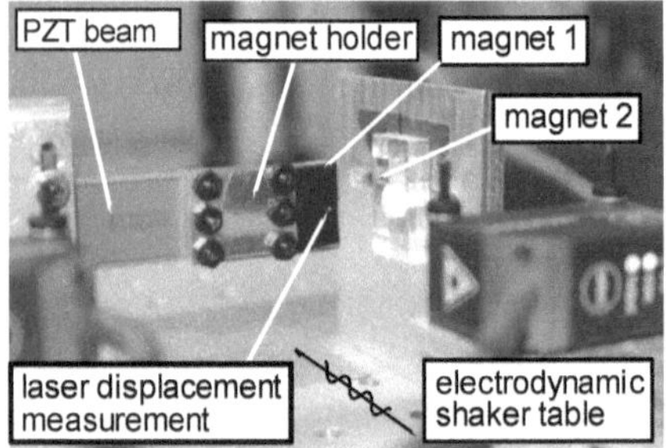


Figure 2.3 Photo of experimental platform.

For the experimental platform in Figure 2.3, classical relations are first employed to determine the lumped mass $m$ and linear stiffness $k$ of the lowest order displacement response [67]. The two static equilibrium positions $x_1$ and $x_2$, and the corresponding natural frequencies $\omega_{n1}$ and $\omega_{n2}$ are identified by impulsive ring-down tests. The viscous damping ratio is then calculated by the logarithmic decrement method. With these values determined and measured, the nonlinear stiffness parameters are calculated by Eq. (2.30).

$$k_2 = \frac{-m\left(\omega_{n1}^2 - \omega_{n2}^2\right)}{x_1 - x_2}; \quad k_3 = \frac{m\left(\omega_{n1}^2 - \omega_{n2}^2\right)}{x_1^2 - x_2^2}; \quad p = 1 + \frac{k_2}{k_1}x_1 + \frac{k_3}{k_1}x_1^2 = 1 + \frac{k_2}{k_1}x_2 + \frac{k_3}{k_1}x_2^2 \qquad (2.30\ \text{a-c})$$

In order to comprehensively examine the influence of noise and asymmetry in the following report, three energy harvesting systems are selected. System 1 is a symmetrical system, and the other two systems have different extents of asymmetry. The identified system parameters are presented in Table 2.2. In the experiments, the resistance is fixed at 100 k$\Omega$ to ensure the circuit working in the optimal condition. The detailed influence of resistance on the rectified power can be found in the references [50] [68].

Table 2.2 Identified system parameters for three nonlinear systems studied in this research.

| | $m$ (g) | $p$ (dim) | $k_1$ (N/m) | $k_2$ (kN/m$^2$) | $k_3$ (MN/m$^3$) | $d$ (N·s/m) | $\alpha$ (mN/V) | $C_p$ (nF) | $R$ (k$\Omega$) |
|---|---|---|---|---|---|---|---|---|---|
| Sys 1 | 18.17 | 1.18 | 556 | 0 | 24 | 0.15 | 1.1 | 88 | 100 |
| Sys 2 | 18.17 | 1.18 | 556 | 4 | 23 | 0.15 | 1.1 | 88 | 100 |
| Sys 3 | 18.17 | 1.16 | 556 | 27 | 26 | 0.15 | 1.1 | 88 | 100 |

## 2.4 Analytical model validation and discussion

The combined harmonic and stochastic excitation is applied to drive the energy harvesting system to validate the analytical model and characterize the electromechanical responses of the three nonlinear system configurations shown in Table 2.2. In the following investigations, the harmonic amplitude contribution to the base acceleration is 3.3 m/s$^2$ at frequencies of 9 Hz for system 1 and 2 and 10 Hz for system 3. The standard deviation of the stochastic base acceleration component is changed over the range of 0 to 1.6 m/s$^2$. To numerically simulate the responses, the fourth-order stochastic Runge-Kutta numerical method [69] is utilized, using 15 normally distributed and randomly selected initial conditions under each specific combination of the harmonic and stochastic base acceleration. The simulation duration is 8000 harmonic periods. In experiments, the duration of measurements is also 8000 harmonic periods. In order to replicate varying initial conditions experimentally, impulsive disturbances may be applied to the beam tip at the start of a given test. The impulses are empirically found to be around 5 kg·mm/s.

With the parameters shown in Table 2.2, each nonlinear system exhibits two stable equilibria. This distinction leads to three classes of vibration: (i) large amplitude snap-through vibration that jumps between two equilibria, (ii) small amplitude intrawell vibration that vibrates around one equilibrium position, and (iii) aperiodic vibration that cannot coexist with the other two types at a certain frequency. Since aperiodic vibration cannot be predicted through the analytical method, this study focuses on the snap-through and intrawell vibration to examine dynamical characteristics resulting from combined harmonic and stochastic excitation. Detailed information about the three vibration types can be found in the reference [8].

### 2.4.1 Dynamical characteristics of the symmetrical nonlinear energy harvesting system

The symmetrical nonlinear system 1 is first considered with the parameter $\Delta = 0$. The analytical and numerical results of the mean value of displacement and the mean-square of rectified voltage are shown in Figures 2.4(a,c) for noise standard deviations from 0 to 1.6 m/s$^2$. At low noise intensity, specifically less than 0.6 m/s$^2$, because the standard deviation is larger than half of the mean value, the numerical results are shown with the original simulation data. Otherwise, the mean value and standard deviation are used to statistically show the numerical results, which holds for all simulation plots throughout this report. As shown in Figures 2.4(a,c), depending on the initial conditions, the system is seen to realize either snap-through or intrawell vibration for small standard deviations of the stochastic base acceleration component. These distinct behaviors are validated experimentally in Figures 2.4(b,d).

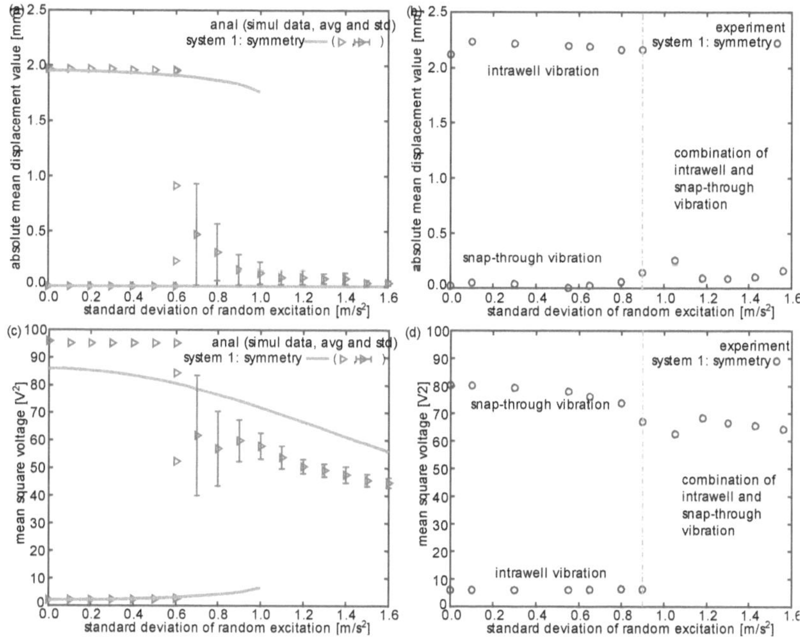


Figure 2.4 Responses for system 1 with harmonic base acceleration amplitude 3.3 m/s² at frequency 9 Hz. Absolute mean value of displacement amplitude from (a) analysis and simulation and (b) experiment. Mean-square of total rectified voltage from (c) analysis and simulation and (d) experiment.

In experiments, the noise standard deviation does not distribute uniformly because of the filter used in generating excitation signals in experiments. In addition, the beam uses a glass-reinforced epoxy layers in the laminate sequence [47]. This leads to inevitable viscoelastic creep so that the mean displacement values increase with time in experiments, especially for the intrawell vibration as shown in Figure 2.5(a). Therefore, the mean displacement values measured from experiments vary when the standard deviation of noise varies and is higher than the mean displacement value calculated from the analysis and simulation. Besides, due to the electrical losses in the rectification circuit that are not accounted for in the models, the measured mean-square rectified voltage is less than analytical and numerical results. Yet, overall, both the qualitative and quantitative range of behaviors observed for the symmetrical nonlinear energy harvesting system excited by the combined harmonic and stochastic base accelerations are in good agreement for lower standard deviations of the noise.

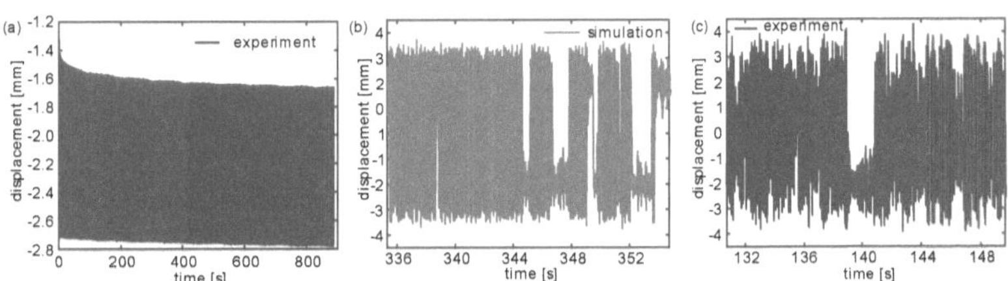

Figure 2.5 Transient displacements from (a) experiment under pure harmonic excitation, (b) simulation and (c) experiment at harmonic excitation amplitude 3.3 m/s$^2$ and noise standard deviation 1.6 m/s$^2$.

Considering the cases for when the standard deviation of the stochastic base acceleration is around 1 m/s$^2$, Figures 2.4(a,c) show that the numerical results indicate regular fluctuations between the snap-through and intrawell vibration levels in terms of the mean displacement and the mean-square rectified voltage. Comparatively, the analysis results in mean-square voltages that appear to follow the average behavior of the simulation trends. Yet, for noise intensity greater than around 1 m/s$^2$, the dependence of the electromechanical responses on the initial conditions is reduced. For much greater noise standard deviation around 1.6 m/s$^2$, the mean values of displacement amplitude in Figure 2.4(a) are close to the value of snap-through vibration and the standard deviation approaches zero. Temporal snap shots of the beam tip displacement are given in Figure 2.5(b) to reveal that the behavior is dominated by snap-through between equilibria, which decreases the mean displacement value in the way shown through Figure 2.4(a). In experiments, Figures 2.4(b,d), when the noise standard deviation is greater than around 1 m/s$^2$, the system only responds with a combination of intrawell and snap-through vibration as seen through the transient response in Figure 2.5(c). The analysis agrees with these trends qualitatively and quantitatively, furthermore revealing good statistical estimates of the mean-square rectified voltage also observed numerically and experimentally.

The trends observed through Figure 2.4 are similar to the influence of noise standard deviation demonstrated in Figure 2.2 that at high noise standard deviation the snap-through vibration happens much more frequently and becomes the dominant vibration type in the responses. In addition, such trends can also be explained through the perspective of potential energy shapes. Since the equivalent linearized model natural frequency $\omega_e$ is in part governed by the stochastic displacement contribution, the stochastic base acceleration has influences on the equivalent potential energy function. Here, Eq. (2.6) is used to determine the potential energy function associated with noise as shown in Figure 2.6. Without noise, a double-well potential energy profile is formed, corresponding to a bistable system configuration. With the noise standard deviation added to 1.6 m/s$^2$, the depth of the potential well shallows to make snap-through vibration happens more frequently. Further increasing the noise standard deviation to 4.8 m/s$^2$, the potential energy function resembles a monostable system with one global minimum, corresponding to a stable equilibrium configuration around which the energy harvester oscillates. This explains the occurrence of snap-through-like vibration with the greater noise components to the base acceleration.

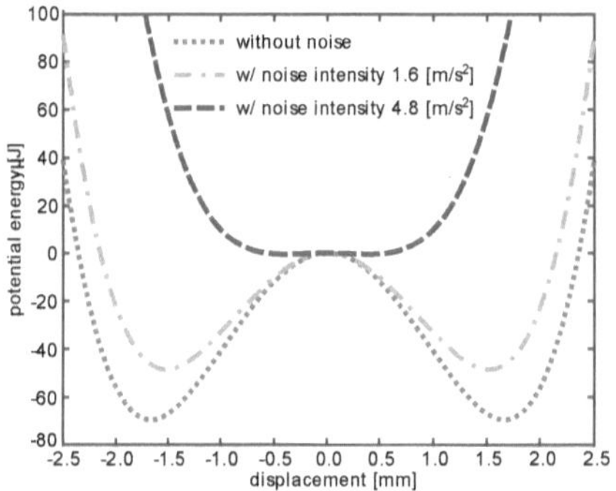


Figure 2.6 Potential energy from linearized governing equation accounting for changes in noise intensity.

### 2.4.2 Dynamical characteristics of asymmetrical nonlinear energy harvesting systems

On the basis of system 1, the magnet 2 is repositioned to set the distance $\Delta$ of 0.02 mm, which is termed system 2. Further moving the magnet 2 to increase the distance $\Delta$ to be around 0.05 mm, the system 3 is constructed. Based on the potential energy plots shown in

Figure 2.7, the potential energy difference between the two potential wells for system 2 is around 20%. Comparatively, for system 3, such difference is increased to 77%. Therefore, in the following discussion, systems 2 and 3 are respectively termed as slight and large asymmetrical systems.

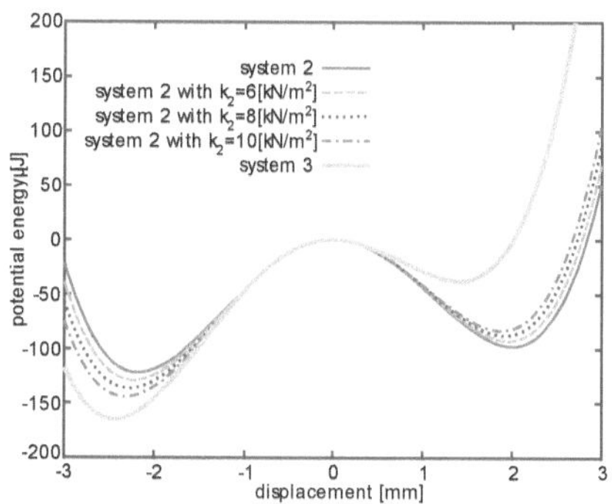


Figure 2.7 Potential energy profiles for the asymmetrical energy harvesting systems.

Considering the range of combined harmonic and stochastic base acceleration, the mean value of displacement and mean-square rectified voltage from analysis and simulation are shown in Figure 2.8(a,c), while the corresponding experimental data is shown in Figure 2.8(b,d). As also observed for the symmetrical system 1, two types of response may either occur for two asymmetrical systems at small noise standard deviation such as less than 0.7 m/s$^2$, as seen in Figure 2.8. Yet, when the noise intensity is above 0.8 m/s$^2$, the existence of noise has different influences on responses for two asymmetrical systems.

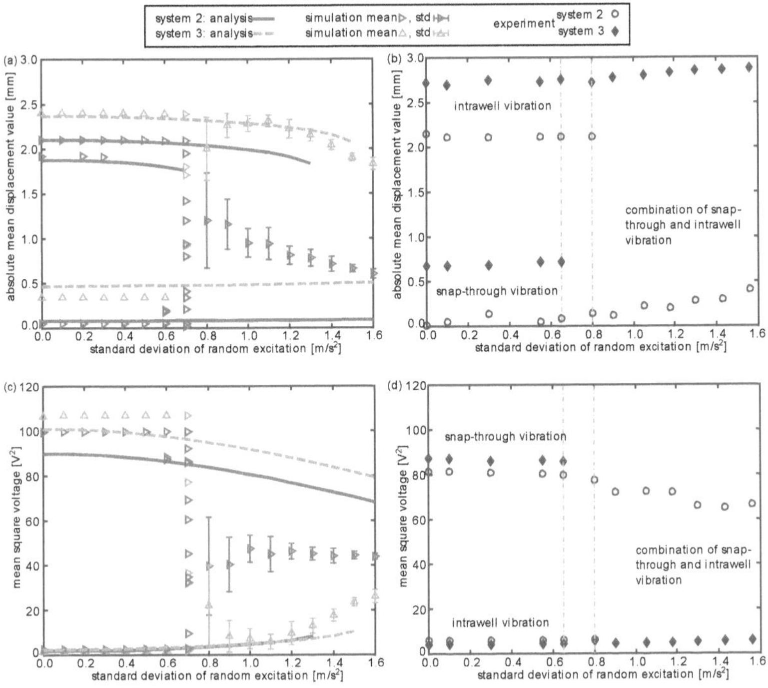


Figure 2.8 Responses for systems 2 and 3 with harmonic amplitude 3.3 m/s² at frequency 9 Hz and 10 Hz. Absolute mean value of displacement amplitude from (a) analysis and simulation and (b) experiment. Mean-square of total rectified voltage from (c) analysis and simulation and (d) experiment.

For slight asymmetrical system 2, similar to the trends demonstrated in Figure 2.4, with the increase of noise intensity, the mean displacement value in simulation is decreased 75% with a near zero standard deviation and the mean-square voltage is the same as in

34

symmetrical system 1 at the noise intensity 1.6 m/s$^2$, which suggests a snap-through dominant vibration exists for slight asymmetrical system 2 under the same excitation condition. In experiments, only the intrawell vibrations associated with the deep potential well side are measured. The experimental results show in Figure 2.8(b,d) validate the snap-through dominant vibration at high noise intensity because of the high mean-rectified voltage and low mean displacement value. In comparison with the simulation and experiments, the analytical results for system 2 demonstrate the same changing trend as in simulations and experiments and predict a good estimation on the mean-square voltage.

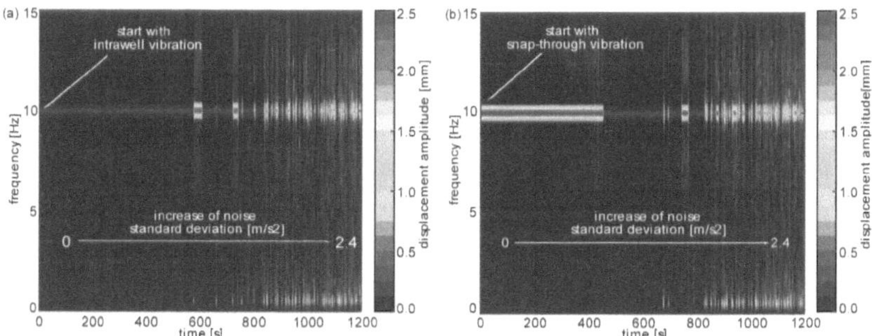


Figure 2.9 Short-time Fourier transform of simulated displacement response across the duration with the increase of noise standard deviation for system 3. Response starts with (a) intrawell vibration and (b) snap-through vibration.

Comparatively, for system 3 with large asymmetry, when under the same excitation condition, the rectified voltage drops substantially in simulations and experiments when the noise intensity is higher than 0.8 m/s$^2$ as in Figure 2.8(c,d), since snap-through vibration amplitudes are not triggered. Here, the mean value of the displacement is large indicating intrawell vibration. This trend is opposite to that of system 1 and 2. The Fourier transforms of the simulated displacement responses for system 3 across the time duration are presented in Figure 2.9. The results in Figure 2.9(a,b) show the broad range of spectral behaviors as the standard deviation of noise increases from 0 to 2.4 m/s$^2$. Between Figure 2.9(a) and (b), the initial conditions are selected to either induce intrawell or snap-through vibration for the simulation start when only harmonic base acceleration occurs. In the time durations near 450 to 600 s, both simulations show small amplitude vibration that suggests an incapability to sustain snap-through vibration. Yet, still further increase of noise standard deviation such as around 1200 s in Figure 2.9 reveals a snap-through-like behavior again. In contrast to the specific phenomena identified for the large asymmetrical system in simulations and experiments, the analysis still indicates a similar snap-through dominant vibration as in systems 1 and 2.

In order to explain the discrepancy between analysis and simulation for greater system asymmetry, three additional cases of asymmetry are considered. System 2 leads to a value for stiffness parameter $k_2$ of 4 kN/m$^2$ . In the additional asymmetrical systems, the parameter $k_2$ is changed to 6 kN/m$^2$, 8 kN/m$^2$, or 10 kN/m$^2$. Figure 2.7 shows the potential energy shapes for the asymmetrical energy harvesting systems. As shown in Figure 2.7, the depth of the potential wells change as a result of the differing asymmetry. With the increase of $k_2$, the difference between two potential wells depth increases.

For the case that snap-through vibration is frequently induced, the probability distribution function accurately employs the factor of 1/2 for the two distribution functions shown in

Eq. (2.24) so long as the system is symmetric so that the depths of the potential wells are identical. Yet, with the introduction of asymmetry, it is not equally likely that the mean-square displacement will adopt a zero mean value. Consequently, an accommodation is required to characterize amount by which the probability distribution function changes. The Eq. (2.31) modifies the original distribution in Eq. (2.24) according to the coefficients $p_1$ and $p_2$ which are the potential energies respectively associated with the stable equilibrium position $x_1$ and $x_2$. In this way, the cumulative contributions from the two distributions $g(x_{rm}|x_1,\sigma_{xg})$ and $g(x_{rm}|x_2,\sigma_{xg})$, which are Gaussian distributions with mean values $x_1$ and $x_2$, account for the asymmetry on an energy basis.

$$f_3(x_{rm}) = \frac{p_1}{p_1 + p_2} g(x_{rm}|x_1,\sigma_{xg}) + \frac{p_2}{p_1 + p_2} g(x_{rm}|x_2,\sigma_{xg}) \qquad (2.31)$$

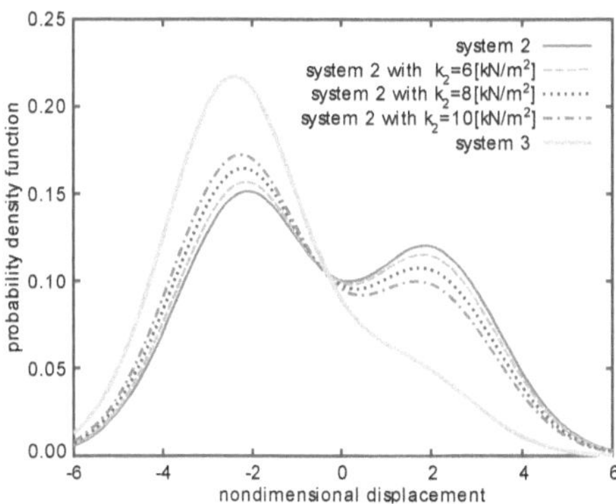


Figure 2.10 Probability density functions for the asymmetrical energy harvesting systems

Using the expression in Eq. (2.31) considering that the radicand in Eq. (2.22b) is a unit valued constant, Figure 2.10 presents the probability density functions for the asymmetrical systems when subjected to non-dimensional noise standard deviation $\sigma = 1.5$ m/s$^2$. The corresponding mean displacement values are numerically integrated through Eq. (2.31) for the system 2 cases shown in Figure 2.10. The mean displacement values are found to be -0.4951 mm for $k_2 = 4$ kN/m$^2$, -0.7351 mm for $k_2 = 6$ kN/m$^2$, -0.8852 mm for $k_2 = 8$ kN/m$^2$, and -1.167 mm for $k_2 = 10$ kN/m$^2$. The dynamical responses for the systems characterized in Figure 2.10 are shown in Figure 2.11. At noise standard deviation $\sigma = 2$ m/s$^2$, the mean displacement value from simulations in Figure 2.11 are around the mean displacement value calculated from the probability functions and listed above. This validates the energy-weighted coefficients used in Eq. (2.31).

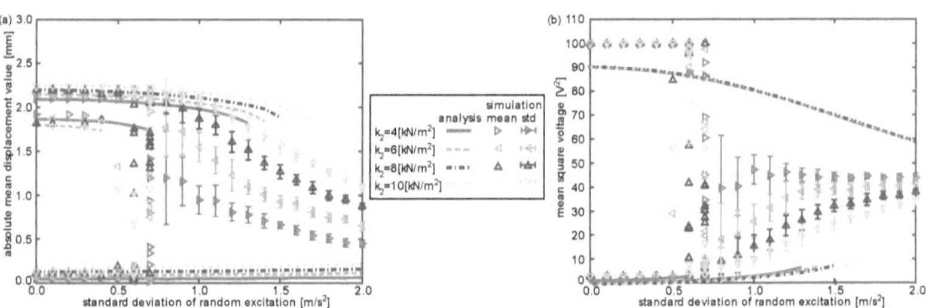


Figure 2.11 Analytical and numerical responses for four asymmetrical energy harvesting systems with harmonic amplitude 3.3 m/s$^2$ and frequency 9 Hz. (a) Absolute mean value of displacement amplitude. (b) Mean square voltage.

Based on the probability density plots in Figure 2.10, for system 3 with large asymmetry there is a much greater possibility to realize intrawell vibration with mean displacements closer to the negative valued equilibrium seen in Figure 2.7. Since the increase in asymmetry may increase the contrast between the weighted Gaussian distribution Eq. (2.23) and the accurate distribution in Eq. (2.31) as in Figures 2.2 and 2.11, the discrepancies in dynamical responses shown in Figures 2.8 and 2.11 may also increase. According to these investigations, when under the same excitation condition, vibration energy harvesting systems exploiting nonlinearities yet subjected to ambient vibrations with harmonic and stochastic contributions are susceptible to reduced performance if the nonlinearity is large enough to cause high asymmetry in the potential energy. For balanced-performance and consistent DC power delivery, it is recommended to reduce asymmetry that may result from nonlinearities in the implementation of energy harvesters in practical vibration environments.

**2.5 Discussion on the dynamic responses**

In Section 2.4, the dynamic responses are investigated using one value for the harmonic base acceleration component. As the potential energy profiles vary with asymmetry such as in Figure 2.7, the kinetic energy needed to overcome potential energy barriers and realize snap-through vibration is changed. This indicates that the combination of harmonic and stochastic base accelerations together crucially determine vibration of the energy harvesting system. This section investigates cases in which the distinct dynamic behaviors transition as a result changes in the balance between harmonic and stochastic acceleration components.

Figure 2.12 presents the responses for the systems 1, 2, and 3. In Figure 2.1, systems 1 and 2 are driven with a harmonic base acceleration of 2.7 m/s², while two cases of system 3 are studied using 5 m/s² or 8 m/s². Comparing the results in Figures 2.4(a) and 2.12(a), with the increase of noise standard deviation $\sigma$ the mean displacement value for system 1 decreases to be a value around zero, analogous to the unstable equilibrium position around which snap-through occurs. Yet, unlike the trend in Figure 2.4(c), for a stochastic base acceleration near 0.7 m/s² in Figure 2.12(b), the mean square voltage substantially drops. This is similar to the trend shown in Figure 2.8(c) for the asymmetric systems and indicates that an intrawell dominant vibration occurs. Such vibration behavior is exemplified in Figure 2.13(a) for noise standard deviation 1.6 m/s². From Figure 2.13(a), although the noise may trigger jumps between the two equilibria, the regularity of generating snap-through vibration is low. A similar trend occurs for system 2 in Figure 2.12(a,b) for the smaller value of harmonic base acceleration 2.7 m/s².

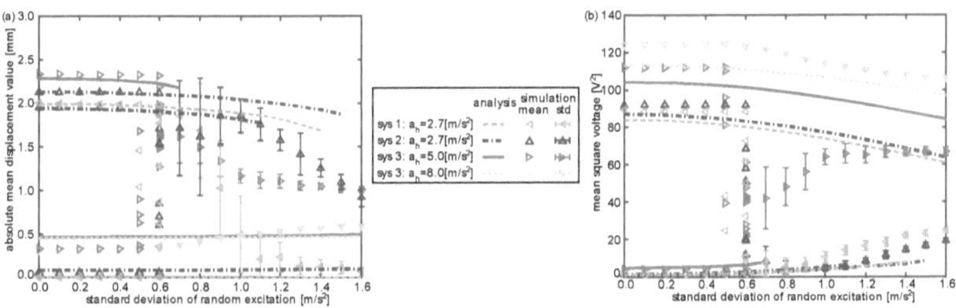

Figure 2.12 Analytical and numerical responses for system 1 and 2 at frequency 9 Hz and system 3 at 10 Hz. (a) Absolute mean value of displacement amplitude. (b) Mean square voltage.

By contrast, for the harmonic base excitation of 5 m/s$^2$, the dynamics of system 3 in Figure 2.12 suggest a snap-through dominant vibration, based on the larger mean square voltage values for most cases of additive noise base acceleration. Figure 2.13(b) shows a time series of the system 3 when driven by 5.0 m/s$^2$ harmonic base acceleration and 1.6 m/s$^2$ standard deviation of stochastic base acceleration. Comparing the analytical and simulation results for the three systems in Figures 2.4, 2.8, and 2.12, for increase of harmonic base excitation amplitude the discrepancy between simulation and analysis decreases. For instance, for system 3 driven by harmonic base acceleration of 8 m/s$^2$, the results of Figure 2.12 suggest that snap-through vibration between the stable equilibria is frequently induced. In such case, the discrepancies between the simulation and analysis are lessened.

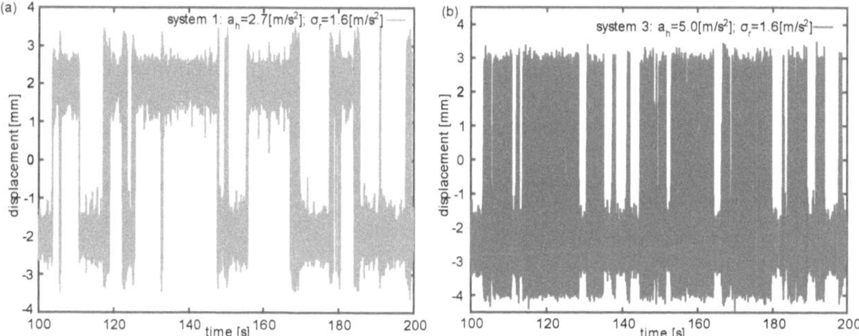


Figure 2.13 Transient displacements from simulation considering stochastic base acceleration of 1.6 m/s$^2$ for (a) system 1 at harmonic excitation 2.7 m/s$^2$ and (b) system 3 at harmonic excitation 5 m/s$^2$.

One explanation for the changes in the discrepancy between analysis and simulation pertains to the influence of the harmonic base acceleration on the effective probability density distribution functions determined by Eq. (2.24) [55]. With increase in the harmonic excitation component, the possibility of triggering the snap-through vibration is increased. For example, when the harmonic base acceleration for system 3 is 8 m/s$^2$, snap-through vibration occurs even without presence of stochastic base acceleration. In other words, once stochastic excitation is then introduced, the probability distribution should be similar to that shown in Figure 2.2(c) regardless of the standard deviation of the noise. In contrast, when the harmonic base acceleration is too small to generate snap-through vibration, the probability density distribution should intuitively appear like that represented in Figure 2.2(a) or Figure 2.10. In such latter case, only with increase of stochastic base acceleration should the effective probability distribution function transition to a unimodal distribution. This suggests that for nonlinear energy harvesting systems subjected to combined harmonic and stochastic base excitation, the weighted Gaussian distribution implementation to lead to Eq. (2.25) results in better estimation of dynamical responses for greater amplitudes of harmonic base acceleration because the response statistics are more comparable to those for a Gaussian distribution [70].

## 2.6 Conclusion

This research examines the dynamics of nonlinear energy harvesting systems under combined harmonic and stochastic excitation to demonstrate the influences of asymmetry and relative balance of excitation components on the electrodynamic responses and DC power delivery. An analytical approach based on the equivalent linearization method is established to account for contributions to dynamics from harmonic and stochastic vibration inputs. A weighted Gaussian joint distribution is adopted in the analytical model to improve the accuracy in predicting the statistical responses due to the stochastic excitation. The discrepancies between simulations and analyses are scrutinized thorough study of the probability density distributions. Combined with validation from simulations and experiments, the study reveals that with the increase of noise intensity, the vibration becomes independent on initial conditions, and snap-through or intrawell dominant vibration may occur. Through the investigations, the findings overall suggest that reducing asymmetry or increasing the harmonic excitation component may trigger snap-through dominant vibration for moderate to high standard deviation of stochastic base acceleration. The outcomes help identify robust energy harvesting system design approaches for consistent DC power delivery when the platforms are subjected to realistic vibration environments.

**Chapter 3 Physics-guided data-driven model for maximizing direct current power**

**of nonlinear energy harvesting systems subjected to periodic impulse excitation**

Vibration energy harvesting is a thoroughly tested approach to support the growing number of Internet-of-Things devices by providing a local and sustainable electrical power resource. Yet, the discovery of best practices for deploying vibration energy harvesters in environments is limited by inability to decipher high-dimensional data associated with design and implementation details. The study aims to devise one such approach to conclusively identify conditions under which an impulse-excited nonlinear energy harvesting system delivers peak electrical power. To accomplish this goal, physics-based and data-driven analyses are integrated to uncover the underlying relationships between the impulse-induced nonlinear dynamics and the result converted electrical energy. The accuracy of the predictions from a machine learning model are confirmed through experiments and against cross-validation simulation data. The parameters that result in snap-through, as determined through data-driven methods, are then probed through a first-principles model to confirm that optimal circuit design conditions agree with the principle of impedance matching. The findings guide attention to previously unseen nuances of exploiting nonlinear vibration energy harvesting systems in impulse excitation environments by revealing parameter sensitivities that inhibit or promote the realization of snap-through vibration for peak power generation. This research demonstrates a successful synthesis of data-driven and first principles models to guide attention to optimal designs of nonlinear energy harvesting systems subjected to periodic impulse excitation.

## 3.1 Introduction

With the ever-growing number of Internet-of-Things (IoT) devices to a trillion by 2025 [39], energy management becomes more of a challenge as conventional chemical batteries have a limited lifespan and undesirable service requirement. Compared with batteries, vibration energy harvesting is a promising option to provide sustainable power from the kinetic energy sources readily available in daily life, such as from human motion, vehicle wheel rotation, bridge vibration, and more [71].

Piezoelectric beams are widely utilized to transfer this kinetic energy to electrical power due to the high-power density of piezoelectric material [40]. Considering the broadband frequency nature that real vibration displays, nonlinearity such as bistability has been introduced to maximize the harvested power in a wide frequency spectrum [72] [15]. Cottone et al. [9] indicated a potential increase of 400% to 600% in harvesting alternate current (AC) power with bistable configuration when subjected to wide-spectrum stochastic excitation. Three possible vibration classes may be generated in bistable structures, including snap-through vibration with displacement crossing the two equilibria, intrawell vibration that oscillates around one equilibrium configuration, and chaotic vibration [8]. Recently, researchers have explored methods to attain high-energy snap-through vibration [47] [11] [73], such as activated potential energy profiles [16], and load perturbations [14]. In addition, since DC power is required to power IoT sensors, nonlinear circuits are investigated when coupled with nonlinear energy harvesters to examine strategies that optimize harvested DC power [74] [75]. Huguet et al. [76] explored the integration of bistable energy harvesters with synchronized electric charge extraction (SECE) circuits and found an increase in electric power and decrease in frequency bandwidth when subjected to harmonic excitation. Cai and Harne [50] proposed an analytical approach to determine optimal working conditions for buck-boost converters interfaced with nonlinear energy harvesters under pure harmonic excitation.

From this survey of state-of-art studies, pure harmonic or stochastic excitation is frequently examined to characterize and optimize the nonlinear dynamics of circuit-harvester integrations for sustainable electric power. Yet, the environment that energy harvesters are likely deployed within may experience complex excitation conditions, for instance impulse excitation induced by human motion, transportation, or other activities [23] [24] [77]. Harne et al. [78] constructed an analytical model with the Jacobian elliptic functions to predict transient electrodynamic behaviors of a bistable energy harvester interfaced with a resistive electric load. Due to the complexity of the Jacobian elliptic functions, the analytical model does not significantly ease the efforts to identify impulse-induced nonlinear dynamics except single impulse events [78] [79]. Instead, numerical methods are widely employed to characterize other impulse-induced responses [80] [81]. Quinn et al. [82] numerically identified improvements in harvested AC power and frequency response for periodically impulse-excited nonlinear energy harvesters. Further, Fang et al. [60] designed an asymmetric plucking-based bistable energy harvester to capture rotational energy by the impulsive event of contact.

While notable advancements have been achieved to integrate nonlinear circuits with nonlinear energy harvesters when the ambient excitations are harmonic or stochastic [68] [74] [75], to the best of the authors' knowledge the interactions among nonlinearities in the harvesting circuits and structures have not been characterized under periodic impulse excitation. Since many impulse-induced vibrations are more repetitive than singular events [23] [24] [77], prior modeling frameworks assume singular impulses on nonlinear energy harvesters do not apply [78] [79]. In addition, researchers [83] have shown that parameters such as impulse magnitude, electromechanical coupling, and viscous damping directly affect the class of vibration that nonlinear energy harvesters may achieve, although only snap-through vibration is desirable for appreciable power generation [12]. Unfortunately, numerical methods are inefficient for the purpose of ad-hoc design optimization since the number of parameters is large, the nonlinearities and dynamics are non-smooth, and periodicity of energy harvester response may not occur in the same time scale as the impulse excitation [83]. Therefore, there remains a need for a unified model that comprehensively characterizes the dynamic mechanisms underlying optimized electrical power extraction from nonlinear energy harvesting systems subjected to periodic impulses. Recent studies have formulated insightful physics-based understanding by harnessing advances in machine learning, especially when theoretical models are extremely challenging to formulate and data-driven methods alone cloud the underlying physics [84] [85] [86]. Brunton et al. [87] identified the governing equations for nonlinear dynamical systems by leveraging sparsity-promoting techniques and machine learning algorithms, demonstrating the ability of such an approach to uncover the salient system dynamics without a requisite theoretical method for guidance. Advancing this concept, Zhang and Sun [88] proposed a physics-guided neural network that employed a loss function to reduce the discrepancy between the neural network model outputs and finite element results. The numerical and experimental validations indicated the strength of the method in improving the generality and scientific consistency of damage detection. Due to the invariance of many systems in fluid dynamics, mechanics, and system dynamics, Ling et al. [89] developed an empirical model by incorporating the invariance properties of simulation data with machine learning processes to demonstrate direct accounting for system symmetries and scalability in the machine learning predictions that are consistent with the underlying physical relations. These examples highlight how emerging advances in data science are accelerating discovery and physical understanding of new problems across disciplines and suggest the broad applicability of such methods to engineering and science.

Inspired by these recent developments and considering the need to illuminate the multiphysics dynamic behaviors of impulse-excited nonlinear energy harvesting systems, this research creates an integrative theoretical and data-driven model for optimizing DC power of nonlinear energy harvesting systems under periodic impulse excitation. Therefore, in the following section a theoretical model of a nonlinear energy harvester subjected to periodic impulses is introduced to study the system dynamics. Following experimental validation of the model, a machine learning approach is described to map the theoretical relations to discrete, non-smooth system dynamics each inducing distinct energy harvesting potential. The integrated machine learning and theoretical models are then explored to characterize dynamic regimes of snap-through vibration and provide guidance for best electrical power extraction. Finally, a summary of key findings is presented to conclude the report.3.2 Nonlinear energy harvesters with magnetic coupling under periodic impulse excitation.

## 3.2 Nonlinear energy harvesters with magnetic coupling under periodic impulse excitation

### 3.2.1 Model for nonlinear energy harvester considering magnet rotation

A piezoelectric beam with additional magnetic potentials is frequently employed to construct a post-buckled energy harvesting system due to the ease of assembly and tuning [90]. Figure 3.1(a) presents one approach to introduce magnetic coupling. In the schematic, a pair of repulsive magnets are employed, one of which is attached to the piezoelectric beam tip with a magnet holder. The position of the other magnet holder is attached to the base, yet remains adjustable in location to regulate the nonlinearities acting on the system by magnetic repulsion. The tip mass $M_0$ at the free tip of the beam is constructed by magnet 1 and the magnet holder. An SEH circuit is connected to the piezoelectric beam to ensure a DC power delivery, Figure 3.1(a).

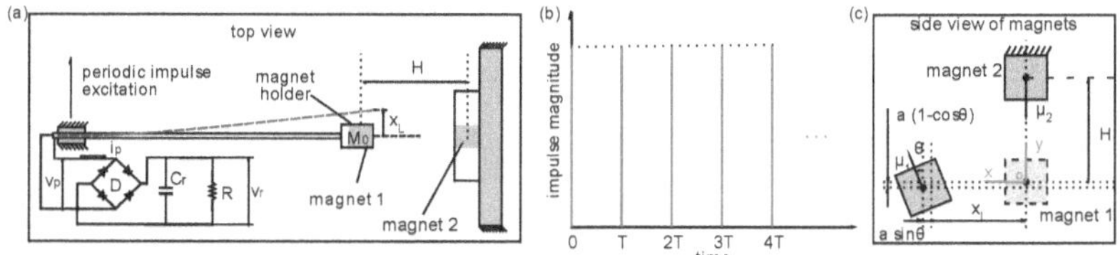


Figure 3.1 (a) Schematic of nonlinear energy harvester connecting with the SEH circuit. (b) Periodic impulse excitation. (c) Detailed schematic of repulsive magnets interaction.

For the nonlinear energy harvesting system shown in Figure 3.1(a), the base (shown in light grey shade) is periodically impulsed, which moves the cantilever beam as well as the holder for magnet 2. Assuming the beam vibrates in the first vibration mode, the governing equation is given in Eq. (3.1).

$$m\ddot{x}_L + c\dot{x}_L + k_1 x_L + k_3 x_L^3 + \alpha v_p = F_{21} \tag{3.1a}$$

$$C_p \dot{v}_p + i_p = \alpha \dot{x}_L \tag{3.1b}$$

$$i_p(t) = \begin{cases} C_r \dot{v}_r + \dfrac{v_r}{R}; \text{ if } v_p = v_r \\ -C_r \dot{v}_r - \dfrac{v_r}{R}; \text{ if } v_p = -v_r \\ 0; \text{ if } |v_p| < v_r \end{cases} \tag{3.1c}$$

In Eq. (3.1), $x_L$ is the beam tip displacement related to the fundamental mode; $m$, $d$, $k_1$, and $k_3$ are the equivalent mass, viscous damping, linear stiffness, and nonlinear stiffness due to the large deflection corresponding to the first vibration mode; $\alpha$ is the electromechanical coupling constant; $C_p$ and $C_r$ are respectively the internal capacitance of the piezoelectric beam and smoothing capacitance; $v_p$ and $v_r$ are respectively the voltage across the piezoelectric beam electrodes and rectified voltage; $i_p$ corresponds to the current passing into the rectification circuit; $F_{21}$ is the magnetic force between magnets 1 and 2; The overdot operator indicates differentiation with respect to time $t$.

Before the application of impulse, the beam is assumed to rest at the static equilibrium position $x^*$. The voltages are across the piezoelectric beam and rectifier are both zero. The initial conditions are given in Eq. (3.2).

$$x_L\big|_{t=0^-} = x^*; \quad \dot{x}_L\big|_{t=0^-} = 0; \quad v_P\big|_{t=0^-} = 0; \quad v_r\big|_{t=0^-} = 0; \qquad\qquad (3.2\text{a-d})$$

Hereafter, the superscript '-' on time indicates the instant before applying the impulse, while the superscript '+' indicates the moment after impulse application.

The periodic impulse excitation applied in the study is assumed to be ideal as shown in Figure 3.1(b). According to the impulse-momentum theorem, the velocity change is chosen to reflect the impulse applied. Therefore, the state changes caused by the periodic impulse excitation are given in Eq. (3.3).

$$x_L\big|_{t=nT^+} = x_L\big|_{t=nT^-} ; \quad \dot{x}_L\big|_{t=nT^+} = \dot{x}_L\big|_{t=nT^-} + I; \quad v_P\big|_{t=nT^+} = v_P\big|_{t=nT^-} ; \quad v_r\big|_{t=nT^+} = v_r\big|_{t=nT^-} ; \qquad n = 0,1,2,3\cdots$$

$$(3.3\text{a-d})$$

The $T$ is the period of the impulse excitation as shown in Figure 3.1(b), $n$ is the number of the impulse cycle that has been applied, and the absolute value of $I$ indicates the magnitude of the impulse exerted to the system. Given the coordinate defined in Figure 3.1(c), the positive value of $I$ represents the impulse applied in the positive direction of the $x$ axis, and the negative value of $I$ suggests the impulse applied in the negative direction.

Since large amplitude vibration is preferred for nonlinear vibration energy harvesting systems, the magnet rotation is necessary to be examined since it determines the nonlinear force acting on the piezoelectric beam tip. For a clamped beam with attached tip-mass, the fundamental mode shape is approximated by

$$W_1(x) = C_4\left[ (\sin\beta_1 x - \sinh\beta_1 x) - \frac{\sin\beta_1 l + \sinh\beta_1 l}{\cos\beta_1 l + \cosh\beta_1 l}(\cos\beta_1 x - \cosh\beta_1 x) \right] \qquad (3.4)$$

where $\beta_1$ corresponds to the first eigenvalue determined by the characteristic equation shown in Eq. (3.5).

$$1 + \frac{1}{\cos \beta_n l \cosh \beta_n l} - R\beta_n l \left( \tan \beta_n l - \tanh \beta_n l \right) = 0 \tag{3.5}$$

In Eq. (3.5), $R = \dfrac{M_0}{\rho A l}$ is a tip mass ratio; $\rho$ and $A$ are the equivalent density and cross section area of beam; and $l$ is the beam length.

Normalizing the mode shape such that the beam tip displacement magnitude is equal to one, the coefficient $C_4$ is determined as shown in Eq. (3.6).

$$C_4 = \frac{1}{\left( \sin \beta_1 l - \sinh \beta_1 l \right) - \left( \sin \beta_1 l + \sinh \beta_1 l \right)\left( \cos \beta_1 l - \cosh \beta_1 l \right) / \left( \cos \beta_1 l + \cosh \beta_1 l \right)} \tag{3.6}$$

Therefore, the rotation angle for the first order model shape is

$$W_1'(x) = C_4 \left[ \beta_1 (\cos \beta_1 x - \cosh \beta_1 x) - \frac{\sin \beta_1 l + \sinh \beta_1 l}{\cos \beta_1 l + \cosh \beta_1 l} \beta_1 \left( -\sin \beta_1 x - \sinh \beta_1 x \right) \right] \tag{3.7}$$

The rotational angle at the beam free tip is then achieved by Eq. (3.8).

$$\theta = W_1'(l) = \frac{(\cos \beta_1 l - \cosh \beta_1 l) - \dfrac{\sin \beta_1 l + \sinh \beta_1 l}{\cos \beta_1 l + \cosh \beta_1 l}\left( -\sin \beta_1 l - \sinh \beta_1 l \right)}{(\sin \beta_1 l - \sinh \beta_1 l) - \dfrac{\sin \beta_1 l + \sinh \beta_1 l}{\cos \beta_1 l + \cosh \beta_1 l}\left( \cos \beta_1 l - \cosh \beta_1 l \right)} \beta_1 x_L = \Psi x_L \tag{3.8}$$

When considering the magnet rotation, the position of magnet 1 with regard to magnet 2 shown in Eq. (3.9) is determined in the coordinate system $(x, y)$ as in Figure 3.1(b), which takes the unstable equilibrium position as the origin.

$$\mathbf{r}_{12} = r_{12,i}\mathbf{i} + r_{12,j}\mathbf{j} = \left[ x_L + a\sin(\theta) \right]\mathbf{i} + \left[ -H - a\left(1 - \cos(\theta)\right) \right]\mathbf{j} \tag{3.9}$$

Here, the characters in bold indicate vectors, $\mathbf{i}$ and $\mathbf{j}$ are unit vectors of the Cartesian coordinate system $(x, y)$, $a$ is the half length of the magnet, and $H$ is the center to center distance between the two magnets.

The magnetic force on magnet 1 induced by magnet 2 is then given in Eq. (3.10) [90].

$$\mathbf{F}_{21} = \frac{3\mu_0}{4\pi\left|\mathbf{r}_{21}\right|^4}\left\{\left(\overline{\mathbf{r}}_{21}\times\boldsymbol{\mu}_2\right)\times\boldsymbol{\mu}_1 + \left(\overline{\mathbf{r}}_{21}\times\boldsymbol{\mu}_1\right)\times\boldsymbol{\mu}_2 - 2\overline{\mathbf{r}}_{21}\left(\boldsymbol{\mu}_2\cdot\boldsymbol{\mu}_1\right) + 5\overline{\mathbf{r}}_{21}\left[\left(\overline{\mathbf{r}}_{21}\times\boldsymbol{\mu}_2\right)\cdot\left(\overline{\mathbf{r}}_{21}\times\boldsymbol{\mu}_1\right)\right]\right\} \quad (3.10)$$

The overbar indicates normalization of the vectors, $\mu_0$ is the free permeability constant, which is $\mu_0 = 4\pi\times10^{-7}\,N/A^2$, $\boldsymbol{\mu}_1$ and $\boldsymbol{\mu}_2$ are respectively the moment of magnet 1 and 2, which are calculated by Eq. (3.11).

$$\boldsymbol{\mu}_1 = -M_1V_1\sin\theta\,\mathbf{i} + M_1V_1\cos\theta\,\mathbf{j}\,;\; \boldsymbol{\mu}_2 = M_2V_2\sin\theta\,\mathbf{i} - M_2V_2\cos\theta\,\mathbf{j} \qquad (3.11\text{a-b})$$

In Eq. (3.11), $M_1$ and $M_2$ are magnetization of magnets 1 and 2; $V_1$ and $V_2$ correspond to the volume of magnets 1 and 2.

For the clamped beam shown in Figure 3.1(a), the magnetic force acting on the magnet 1 along the transverse $x$ direction is computed to be

$$F_{21} = \frac{3\mu_0 M_1V_1 M_2V_2}{4\pi}\frac{-5r_{12,i}\left(r_{12,i}\cos\theta + r_{12,j}\sin\theta\right)r_{12,i} + \left(4r_{12,i}\cos\theta + r_{12,j}\sin\theta\right)\left(r_{12,i}^2 + r_{12,j}^2\right)}{\left(r_{12,i}^2 + r_{12,j}^2\right)^{7/2}} \quad (3.12)$$

Substituting the expression of magnet force Eq. (3.12) to the governing equation in Eq. (3.1) with the initial conditions Eq. (3.2) and periodic impulsive excitation Eq. (3.3), the nonlinear system dynamics of the energy harvesting system are able to be examined.

### 3.2.2 Discussions on dynamics of nonlinear energy harvesters under periodic impulse excitation

Given the model established in Section2.1, two systems are selected to investigate the dynamics of the nonlinear energy harvester subjected to periodic impulse excitation. The parameters used in system 1 are shown in Table 3.1. The system parameters selected for system 1 induce bistability and thus result in considerably distinguished dynamics between low and large amplitude vibrations. The magnet distance for system 1 is decreased by 5% to construct system 2. Thus, system 2 possesses stable equilibria that are further spaced apart due to the greater repulsive force. The frequency of the periodic impulse in Figure 3.1(b) is hereafter referred to as the periodic impulse frequency. In the following investigations, the periodic impulse frequency is changed over the range of 6 Hz to 15 Hz. The impulse magnitudes are selected to be 0.06 m/s and 0.075 m/s for systems 1 and 2, respectively, which are sufficient to trigger snap-through vibration over at least a portion of the frequency spectrum. Two impulse directions are taken into account. Specifically, the impulse direction with negative impulse velocity is defined as the negative direction. Correspondingly, the opposite direction is specified as a positive impulse direction. The fourth-order Runge-Kutta numerical method is utilized to determine the responses from Eq. (3.1). The initial conditions are given in Eq. (3.2). The simulation duration is chosen to be 500 impulsive periods to ensure the steady-state responses are identified.

Figure 3.2 investigates the characteristics of the dynamic responses when the nonlinear energy harvesting system is subjected to periodic impulse excitation. As shown in Figure 3.2(a), similar to the frequency responses under harmonic excitation [8], for lower periodic impulse frequencies the small amplitude intrawell vibration is induced for both systems. With the increase of the periodic impulse frequency to around 9 Hz, there is a discrete increase in the displacement amplitude by almost 10 times, illustrating the generation of snap-through vibration. For the periodic impulse frequency greater than 12 Hz, the intrawell vibration is induced for both systems. The decrease in magnet distance in system 2 compared to system 1 leads to a higher nonlinearity, which corresponds to a larger static equilibrium value. Since the snap-through vibration jumps between two static equilibrium

positions, the displacement amplitude generated for system 2 is more than those amplitudes

induced for system 1 at the same periodic impulse frequencies.

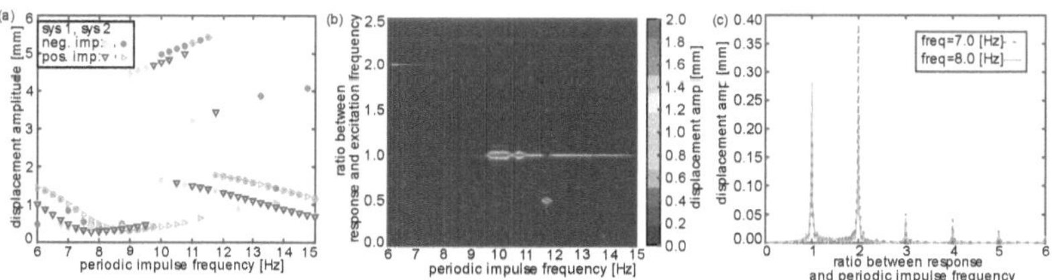

Figure 3.2 (a) Frequency responses of beam tip displacement amplitude for systems 1 and

2 having two magnet gaps dimensions and negative and positive directions of the periodic

impulse. (b) Fourier transform of displacement amplitude across excitation frequencies for

system 1. (c) Fourier transform of displacement amplitude at frequencies 7.5 Hz and 8 Hz

for system 1.

Moreover, according to the Fourier transform of system 1 displacement in Figure 3.2(b),

the response at each periodic impulse frequency is dominated by a single frequency, which

is an integer multiple of the periodic impulse frequency. In the following discussions, the

ratio of the dominant frequency of the response and the periodic impulse frequency is

termed the dominant frequency ratio. As seen in Figure 3.2(b), the dominant frequency

ratio shifts from two to one when the periodic impulse frequency is changed from 6 Hz to

10 Hz. The transition occurs around 8 Hz. As shown in the frequency responses for system

1 in Figure 3.2(c), the dominant frequency ratio is two at a periodic impulse frequency of

7 Hz and changes to a ratio of one at 8 Hz. In addition, around the periodic impulse frequency that the transition happens, the displacement amplitude at the dominant frequency ratio is slightly greater than the second-largest displacement amplitude, which implies a combined frequency influence in the responses.

One noticeable contrast between responses generated from positive and negative impulses for two systems lies in the high-energy snap-through vibration. As shown in Figure 3.2(a), in terms of snap-through vibration, the negative impulse can more often generate the snap-through vibration at lower frequencies for both systems. Figures 3.3(a,b) respectively show the time series of displacement and velocity for system 2 at the periodic impulse frequency 9 Hz with two impulse directions. At time $t = 0^+$, the application of the negative impulse introduces a negative velocity in the system. According to the initial conditions shown in Eq. (3.2) and the changes of states in Eq. (3.3), the negative impulse here is defined as $I_n = -42$ mm/s. The negative sign '-' indicates the impulse is applied towards the negative $x$ axis as in Figure 3.1(c). Comparatively, the positive impulse is $I_p = 42$ mm/s exerts an impulse in the positive $x$ axis.

Figure 3.3(c) displays the potential energy of the nonlinear system and total energy of the time series derived from Figure 3.3(a). It can be seen from Figure 3.3(c) that the potential energy around the positive static equilibrium is asymmetric. As such, the magnitudes of the beam tip velocity generated by the two impulse directions are distinct at every moment except for times $t = 0$ and $t = T_n$ in Figures 3.3(a,b). Here $T_n$ corresponds to the period of the beam's free vibration. As shown in Figure 3.3(b), at the instant before the application of the impulse at $t = T$, the values of the velocities associated with the two impulse directions are negative, which indicates that the beam velocity is in the negative direction of the $x$ axis. Based on the change of velocity caused by the impulse shown in Eq. (3.3), the application of the negative impulse increases the magnitude of the beam velocity by 42 mm/s. Yet, the velocity magnitude associated with the positive impulse direction decreases in Figure 3.3(b). Thus, two opposite effects may be induced by the two impulse directions.

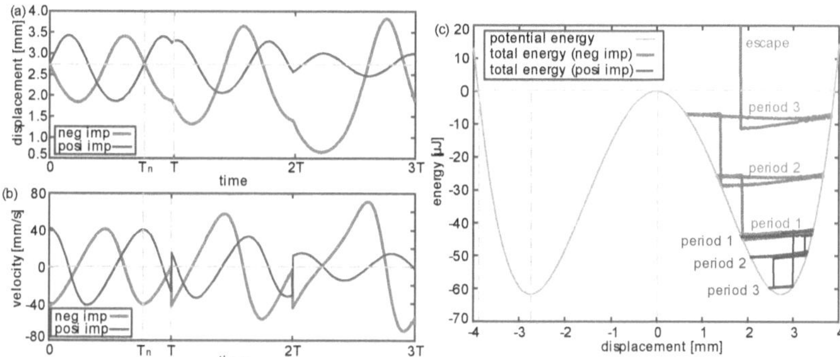


Figure 3.3 (a) Displacement time series and (b) velocity time series with magnet gap $H_1$ at periodic impulse 9 Hz and impulse magnitude 42 mm/s in two impulse directions. (c) Potential energy plot for system 2 and total energy changes in first three periods for system 2 at periodic impulse 9 Hz and impulse magnitude 42 mm/s in two impulse directions.

The total energy changes for the vibration shown in Figure 3.3(a) are presented in Figure 3.3(c) as functions of the displacement to permit contrast with the potential energy profile. For the response induced by the negative impulse direction, the application of the impulse keeps injecting kinetic energy into the system. The accumulated energy helps the beam

57

escape the potential energy barrier and start generating snap-through vibration after 3 cycles of excitation. In contrast, the kinetic energy for the system excited by the positive impulse decreases in time, seen by the confinement of the displacement to positions within one well of the potential energy profile. In addition, because the magnitudes of the displacement and velocity at $t = T_n$ are the same for the two impulse directions. When the periodic impulse frequency is close to the natural frequency of the vibration system $1/T_n$, the velocities generated by the two impulse directions are around same magnitude and in opposite direction. As such, both impulse directions may generate snap-through vibration, such as frequencies around 10 Hz for system 1 and frequencies around 11 Hz for system 2 in Figure 3.2(a).

According to these preliminary observations, both high amplitude snap-through and low amplitude intrawell vibrations may be generated with a dominant frequency component by the periodic impulse excitation. The distinct dynamic regimes have significant implications on energy harvesting effectiveness. In the following sections, these findings are utilized to construct a reduced order surrogate model to improve one's ability to design and analyze the dynamics of the nonlinear energy harvesting systems.

## 3.3 Experimental validation

### 3.3.1 Experimental system description and identification

To experimentally validate the key findings described above, the platform shown in the schematic of Figure 3.1(a) is realized in the laboratory as seen in Figure 3.4. Figure 3.4(a) presents the piezomagnetoelastic system studied here. The piezoelectric beam is a bimorph made by sequential layers of FR4, PZT-5H, and steel shim. The PZT-5H layer covers part of the beam for sake of minimizing the brittle PZT usage. The beam is clamped to an aluminum base, which is affixed to the mounting table. A permanent magnet is secured in a magnet holder and is attached to the free end of the beam. The total tip mass $M_0$ is the combination of the magnet and magnet holder. The other magnet is mounted adjacent to the beam free end, and is oriented so as to introduce repulsive magnet force upon the beam tip. Two displacement laser sensors (Micro-Epsilon ILD-1420) are utilized to measure the absolute displacements of the shaker table and the beam tip. A standard rectifier bridge (1N4148 diodes) is connected to the piezoelectric beam to convert an AC signal to a DC signal. Following the bridge, a smoothing capacitor $C_r$ and resistive load $R$ are utilized to quantify the harvested electric energy.

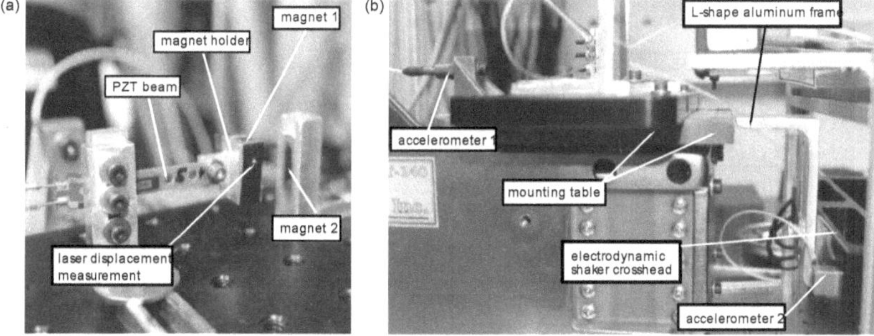


Figure 3.4 Photos of experimental setup. (a) Nonlinear energy harvester. (b) Equipment to generate periodic impulse excitation.

Figure 3.4(b) shows the equipment to generate the periodic impulse excitation. In Figure 3.4(b), the mounting table is fixed on a base that can provide necessary stiffness and damping. The mounting table is struck by an L-shape aluminum frame. The L-shape aluminum frame is installed at the crosshead of an electrodynamic shaker table (APS Dynamics 400). The gap between the L-shape aluminum frame and the mounting table is maintained at 2 mm before the beginning of each experiment. The electrodynamic shaker table is driven by a controlled harmonic signal provided by a controller (Vibration Research Controller VR9500) and amplifier (Crown XLS 2500). An accelerometer (PCB Piezotronics 333B40) is utilized to collect the feedback signal for the controller. The periodic impulse base acceleration on the mounting table is generated by the impact from the aluminum frame striking the mounting table. The opposite impulse direction is achieved by rotating the mounting table by 180°. The accelerometer (PCB Piezotronics 352C04) is installed to measure the frequency and magnitude of the impulse acceleration.

Table 3.1 Identified system parameters for the nonlinear system.

| $m$ (g) | $c$ (N·s/m) | $k_1$ (N/m) | $k_3$ (MN/m$^3$) | $\rho$ (kg/m$^3$) | $A$ (mm$^2$) | $l$ (mm) | $a$ (mm) |
|---|---|---|---|---|---|---|---|
| 10.6 | 0.01 | 213 | 2 | 3500 | 7.5 | 34 | 3.175 |
| $C_p$ (nF) | $\alpha$ (mN/V) | $H$ (mm) | $C_r$ ($\mu$F) | $M$ (MA/m) | $V$ (cm$^3$) | $R$ (k$\Omega$) | |
| 44 | 0.07 | 21 | 47 | 1.25 | 0.768 | 560 | |

For the piezomagnetoelastic system, the equivalent lumped mass $m$ and linear stiffness $k_1$ are determined according to the classical relations [67]. The logarithmic decrement method is then applied to determine the viscous damping. With the measured static equilibrium position $x^*$ and magnet distance $H$, the nonlinear term $k_3$ is determined by Eq. (3.13).

$$k_1 x^* + k_3 \left(x^*\right)^3 = F_{21}^* \tag{3.13}$$

The parameters of the nonlinear energy harvester platform examined in the laboratory are provided in Table 3.1.

### 3.3.2 Model validation and discussions

In order to validate the efficacy of the model established in Section 3.2, the experimental setup shown in Figure 3.4 is utilized. The influences of impulse direction on the generation of snap-through vibrations and frequency dominant responses are closely considered in experiments. Five frequencies are selected to excite the system with six impulse magnitudes in two impulse directions.

Table 3.2 shows the experimental and numerical results that indicate the vibration type and the dominant frequency component. The impulse magnitude characterized by the velocity change is integrated from the impulse acceleration acquired from experiments. The signs '+' and '-' indicate the impulse direction, while vibration type 1 is intrawell and type 0 is snap-through vibration. In experiments, intrawell vibrations result for periodic impulse frequencies from 6 Hz to 10 Hz at lower impulse magnitudes, such as 52 mm/s at 9 Hz. With the increase of the impulse magnitude to 59 mm/s, snap-through vibrations are then generated at 9 Hz for the impulse in the negative $x$ direction. Further increasing the impulse magnitude to 100 mm/s in the negative $x$ direction, snap-through vibrations can be realized at 6 Hz. Comparing the snap-through vibrations generated by the two impulse directions, the same impulse magnitude with the negative direction is seen to more often lead to snap-through vibrations, which is highlighted by the light grey background in Table 3.2. For example, for periodic impulse frequency of 8 Hz and impulse magnitude greater than 68 mm/s, the impulses with negative direction always lead to snap-through vibrations. In comparison, positive impulses generate intrawell vibrations. At 9 Hz, both impulse directions generate snap-through vibrations when the impulse magnitude is greater than 64 mm/s, which indicates a decreased influence of impulse directions on snap-through vibration at higher periodic impulse frequency as discussed in Section 3.2.2.

Table 3.2 Comparison between experiments and simulations. Impulse demonstrates impulse magnitude and impulse direction. The positive sign '+' indicates positive impulse direction and '-' indicates negative impulse direction. Vibration type 1 represents intrawell vibration, 0 represents snap-through vibration. Frequency ratio is the dominant frequency ratio, determined from the ratio between response dominant frequency and periodic impulse frequency. Light grey shading indicates differences in generating snap-through vibration by two impulse directions. Dark grey shading indicates discrepancies between simulations and experiments in predicting vibration type and frequency ratio.

| $f_{exc}$ [Hz] | Classification | Experiment | | | | | | | | | | | | Simulation | | | | | | | | | | | |
|---|---|---|---|---|---|---|---|---|---|---|---|---|---|---|---|---|---|---|---|---|---|---|---|---|---|
| | Impulse $I$ | 67 | | 79 | | 90 | | 100 | | 113 | | 133 | | 67 | | 79 | | 90 | | 100 | | 113 | | 133 | |
| 6 | [mm/s] | + | - | + | - | + | - | + | - | + | - | + | - | + | - | + | - | + | - | + | - | + | - | + | - |
| | Vib. type | 1 | 1 | 1 | 1 | 1 | 1 | 1 | 0 | 0 | 0 | 0 | 0 | 1 | 1 | 1 | 1 | 1 | 1 | 0 | 0 | 0 | 0 | 0 | 0 |
| | Freq. ratio | 2 | 2 | 2 | 2 | 2 | 2 | 2 | $\frac{3}{2}$ | $\frac{3}{2}$ | $\frac{3}{2}$ | $\frac{3}{2}$ | $\frac{3}{2}$ | 2 | 2 | 2 | 2 | 2 | 2 | 2 | 2 | 2 | 2 | 2 | 2 |

Continued

Table 3.2 Continued

| No. | | | | | | | | | | | | | | | | | | | | | | | | | | |
|---|---|---|---|---|---|---|---|---|---|---|---|---|---|---|---|---|---|---|---|---|---|---|---|---|---|---|
| 7 | Impulse | 67 | | 88 | | 107 | | 114 | | 118 | | 130 | | 67 | | 88 | | 107 | | 114 | | 118 | | 130 | |
| | $I$ [mm/s] | + | - | + | - | + | - | + | - | + | - | + | - | + | - | + | - | + | - | + | - | + | - | + | - |
| | Vib. type | 1 | 1 | 1 | 1 | 1 | 1 | 1 | 1 | 1 | 0 | 1 | 0 | 1 | 1 | 1 | 1 | 1 | 1 | 1 | 1 | 0 | 0 | 0 | 0 |
| | Freq. ratio | 2 | 2 | 2 | 2 | 2 | 2 | 2 | 2 | 2 | 1 | 2 | 1 | 2 | 2 | 2 | 2 | 2 | 2 | 2 | 2 | 0 | 0 | 2 | 2 |
| 8 | Impulse | 54 | | 60 | | 65 | | 68 | | 76 | | 85 | | 54 | | 60 | | 65 | | 68 | | 76 | | 85 | |
| | $I$ [mm/s] | + | - | + | - | + | - | + | - | + | - | + | - | + | - | + | - | + | - | + | - | + | - | + | - |
| | Vib. type | 1 | 1 | 1 | 1 | 1 | 1 | 1 | 0 | 1 | 0 | 1 | 0 | 1 | 1 | 1 | 1 | 1 | 1 | 1 | 1 | 1 | 1 | 1 | 1 |
| | Freq. ratio | 2 | 2 | 2 | 2 | 2 | 2 | 2 | 1 | 2 | 1 | 2 | 1 | 1 | 1 | 1 | 1 | 1 | 1 | 1 | 1 | 1 | 1 | $\frac{3}{2}$ | 1 |
| 9 | Impulse | 30 | | 46 | | 52 | | 59 | | 64 | | 67 | | 30 | | 46 | | 52 | | 59 | | 64 | | 67 | |
| | $I$ [mm/s] | + | - | + | - | + | - | + | - | + | - | + | - | + | - | + | - | + | - | + | - | + | - | + | - |
| | Vib. type | 1 | 1 | 1 | 1 | 1 | 1 | 1 | 0 | 0 | 0 | 0 | 0 | 1 | 1 | 1 | 1 | 1 | 1 | 1 | 1 | 1 | 0 | 1 | 0 |
| | Freq. ratio | 2 | 2 | 1 | 2 | 1 | 2 | 1 | 1 | 1 | 1 | 1 | 1 | 1 | 1 | 1 | 1 | 1 | 1 | 1 | 1 | 1 | 1 | 1 | 1 |
| 10 | Impulse | 27 | | 36 | | 54 | | 66 | | 85 | | | | 27 | | 36 | | 54 | | 66 | | 85 | | | |
| | $I$ [mm/s] | + | - | + | - | + | - | + | - | + | - | | | + | - | + | - | + | - | + | - | + | - | | |
| | Vib. type | 1 | 1 | 1 | 1 | 1 | 1 | 1 | 1 | 1 | 1 | | | 1 | 1 | 1 | 1 | 1 | 1 | 1 | 0 | 0 | 0 | | |
| | Freq. ratio | 1 | 1 | 1 | 1 | 1 | 1 | 1 | 1 | 1 | 1 | | | 1 | 1 | 1 | 1 | 1 | 1 | 1 | 1 | 1 | 1 | | |

Further comparing the simulations with experiments in Table 3.2, the numerical model is less accurate in reproducing experimental behaviors for large magnitudes of the impulse. The model predictions that disagree with the experiments are highlighted with dark shaded background in Table 3.2. As shown in the comparison of simulated and measured displacement time series in Figure 3.5(a), the system realizes the intrawell vibration with periodic impulses at 10 Hz and 54 mm/s magnitude, and good agreement is seen between experiment and simulation. Yet, when the impulse magnitude is great enough to generate snap-through vibration, such as 68 mm/s at 9 Hz in Figure 3.5(b), the beam slightly crosses the unstable equilibrium in the experiment whereas numerical responses illustrate a more symmetric snap-through vibration between two stable equilibria. The use of reinforcing FR4 layers in the piezoelectric beam may cause the discrepancies for large displacement amplitude response since FR4 is a viscoelastic fiber-reinforced polymer with nonlinear damping and stiffness that is challenging to characterize in full [91] [92]. The present model assumes linear damping and linear elasticity, which may not completely capture such dynamic behaviors in regimes of large strains and strain rates associated with the largest amplitude of snap-through. Another factor that may cause the discrepancies between experiment and simulation is the non-ideal duration of the impulse. In experiments, the impulse occurs over around 8 ms (around 8% of the excitation period) which is a small but finite proportion of the excitation conditions. Yet, despite these distinguishing points between the model results and experimental findings, agreement is obtained for the great majority of cases studied in Table 3.2, validating the composition of the theoretical model for the many cases of interest considered here.

In terms of the dominant frequency component in response, the dominant frequency ratio shifts from two to one near the periodic impulse frequency 9 Hz in experiments in contrast to 8 Hz in simulations. As shown in Figure 3.5(c), the dominant frequency ratio is two when the periodic impulse frequency and impulse magnitude are 8 Hz and 85 mm/s in the positive direction respectively. A transition occurs to a dominant frequency ratio of one at periodic impulse frequency 9 Hz and impulse magnitude 59 mm/s in the positive direction.

Similar to Figure 3.2(c), the displacement amplitude at the dominant frequency component is slightly greater than the second-largest displacement amplitude at periodic impulse frequency 9 Hz. Since the numerical model adopts assumptions such as lumped parameters and viscous damping to characterize the system, the small variation in displacement amplitude at dominant frequency ratios one and two may not be completely captured. Given the definition on the response dominant frequency, the discrepancy in capturing the displacement amplitude may result in the variations in predicting the dominant frequency ratio at periodic impulse frequencies 8 and 9 Hz in simulations.

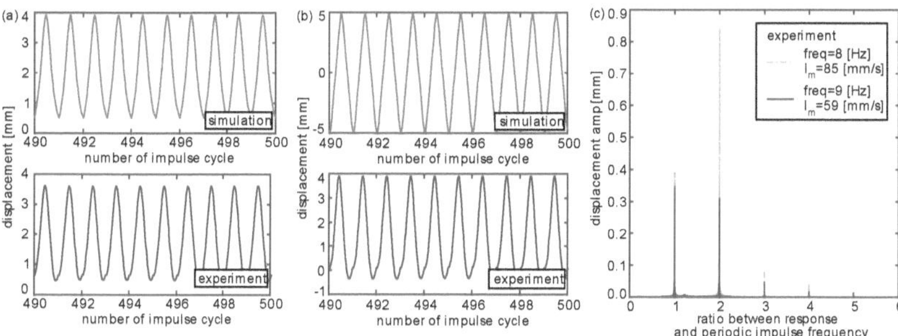


Figure 3.5 Time series of displacement (a) at periodic impulse frequency 10 Hz and impulse magnitude 0.054 m/s in the positive x direction, and (b) at periodic impulse frequency 9 Hz and impulse magnitude 0.067 m/s in the negative x direction. Simulation results are shown at top and experimental results at bottom. (c) Fourier transform of experimentally measured displacement amplitude at frequency 8 Hz with impulse magnitude 85 mm/s in the positive x direction and frequency 9 Hz with impulse magnitude 59 mm/s in the positive x direction.

The optimal resistance for DC power collection estimated by the model is then confirmed. Table 3.3 shows the optimal resistances from experiments and simulations. In experiments, five cases are selected to assess agreement across all vibration types and ratios of the dominant response frequency. Compared with the experiments, the simulations provide good approximations on optimal resistances for frequencies from 6 Hz to 10 Hz, although error increases near periodic impulse excitations around 8 Hz where transitions occur in the values of the dominant frequency ratio (see Table 3.2). Since the response frequency influences optimal resistance selections [50], for excitation frequencies around 8 Hz the differences of optimal resistance between simulations and experiments are within 25%.

Table 3.3 Comparison between experiment and simulation for optimal resistance

| Periodic impulse frequency [Hz] | Impulse $I$ [mm/s] | Optimal resistance [k$\Omega$] | |
|---|---|---|---|
| | | Experiment | Simulation |
| 6 | 79 | 470 | 475 |
| 7 | 67 | 560 | 409 |
| 8 | 54 | 470 | 356 |
| 9 | 67 | 509 | 625 |
| 10 | 54 | 560 | 596 |

Overall, the experimental setup provides means to validate and classify the simulations of nonlinear energy harvesting system dynamics when subjected to periodic impulses. This foundation is essential for subsequent couple the modeling approach into a data-driven prediction framework that helps create insight on the relative merits on the dynamic regimes for vibration energy harvesting.

## 3.4 Physics-guided data-driven model for optimal design strategies

From the perspectives of system design, individual simulations are an inefficient way to identify the optimality by gradient descent methods. Furthermore, the non-smooth and discretized nature of the dynamic behaviors encourage a classification approach that distinguishes parametric combinations for the contributions to best performing system configurations. In this section, a machine learning algorithm is integrated with the theoretical model to assist in identifying mechanisms and design practices that lead to maximum DC power in impulse-excited nonlinear vibration energy harvesting systems.

### 3.4.1 Physics-guided data-driven model for dynamics characterization

#### *3.4.1.1 Establishment of a physics-guided data-driven model*

For high-dimensional and multiphysics problems in science and engineering, reduced-order models (ROMs) are employed to minimize computational needs for only those dynamic behaviors that culminate in the great proportion of meaningful system responses. The ROMs are established by principal components analysis or dynamic mode decomposition to project the intricate governing equations or responses to a low-dimensional and often-linear subspace. Yet, the prediction accuracy may be limited as a consequence of simplification. With the emergent capabilities in data science, machine learning algorithms integrated with theoretical models aim to construct ROMs that are more computationally efficient as well as better low-dimensional representations of an intricate system dynamics [93]. One approach to constructing the physics-guided data-driven ROMs is to build a data-driven surrogate model for a full-order model [93]. For example, Chen et al. [94] exploited the Support Vector Machine algorithm to construct a surrogate reduced-order model to predict the limit cycle oscillation of nonlinear aeroelastic systems. In the following discussions, supervised learning is selected to help build a surrogate model that maps the input design parameters to output dynamics characterized by the vibration type and the dominant frequency ratio.

There are several classification algorithms in supervised learning , such as logistic regression, decision trees, support vector machines, and neural networks. Compared with other methods, neural networks infer the nonlinear and complex relationships among states and parameters with less manual intervention to craft suitable basis functions [95] [96]. Therefore, the neural network algorithm is selected to help build the mapping functions for the surrogate model of the impulse-excited nonlinear vibration energy harvesting system. The network architecture utilized in the study is a three-layer neural network as shown in Figure 3.6(a), which includes an input layer, one hidden layer, and output layer. The units on the input layer are defined by the features of the training examples. Features are design parameters and other factors that are initially assumed to be correlated with outputs. The hidden units create the nonlinear mapping between the input and output layers. The number of units in the hidden layer is chosen to ensure a high accuracy of the prediction results. The output units identify the classes of a problem.

In this research, the inputs corresponds to design parameters of the nonlinear energy harvesting systems, while the outputs are the classes characterized by the vibration type and the dominant frequency ratio. In this way, we seek to correlate the system dynamic behaviors associated with intrawell and snap-through vibration with the periodicity of the responses resulting from the excitation and design parameter combinations. Specifically, four design parameters are selected, which are impulse magnitude $I$, periodic impulse frequency $f_{exc}$, magnet gap $H$, and damping constant $c$. The input features for the neural network is shown in Eq. (3.14). For the given problem, a hidden layer with 25 units is tested to provide satisfying prediction results.

$$\mathbf{x} = \begin{bmatrix} \bar{f}_{exc} & \bar{I} & \bar{H} & \bar{c} \end{bmatrix} \tag{3.14}$$

In Eq. (3.14), the overbar indicates the normalization of features to ensure each feature is on a similar scale. Here the normalization is carried out by subtracting the minimum value and dividing by the maximum value for each respective feature to scale the values from zero to one.

The established model in Section 3.2.1 is utilized to generate the instances for training and cross validating the neural network. In simulations, the impulse $I$ varies from -0.18 m/s to 0.18 m/s to consider the impulse applied in two directions defined in Section 3.2, the periodic impulse frequency $f_{exc}$ ranges from 6 Hz to 15 Hz, the magnet gap $H$ is selected from 19.55 mm to 21.55 mm, and the damping $c$ increases from 0.005 N·s/m to 0.05 N·s/m. The increments assigned for the four parameters are 0.02 m/s, 0.5 Hz, 0.01 mm, and 0.005 N·s/m respectively. In total, 75,810 simulation examples are generated.

Before training the neural network, the numbers of the occurrence for each class are counted. After ordering the occurrence number from high to low, the accumulated summation is calculated. Once the summation of the examples is higher than 95% of the total examples, the classes with occurrence less than 5% are deleted. After the removal of classes with such insignificant probability of occurrence, by changing the 4 input parameters, there remain 68,321 simulation examples that are classified into 6 distinct classes. Here class 1 refers to intrawell responses with dominant frequency ratio two; class 2 refers to intrawell responses with dominant frequency ratio one; class 3 refers to snap-through vibration with dominant frequency ratio one; class 4 refers to intrawell vibration with dominant frequency ratio three; class 5 refers to chaotic vibration; and class 6 refers to intrawell vibration with a dominant frequency ratio of four. Figure 3.6(b) shows the numerical classification results by changing the periodic impulse frequency from 6 Hz to 15 Hz and impulse magnitude from 0 to 0.18 m/s in two directions with the values of magnet gap as 20.55 mm and damping as 0.01 N.s/m. With these values of magnet gap and viscous damping, there are no examples of class 6 in this example dataset. The classes shown in rectangles in Figure 3.6(b) belong to 6 additional classes that have insignificant probabilities to occur, such as intrawell vibration with dominant frequency ratio of one-half. According to Figure 3.6(b), the remaining examples exemplify the predominant dynamic characteristics of the system.

With the remaining simulation examples, 70% of the examples are selected to construct the training set, the remaining 30% are taken as the cross-validation set. For the training set, as shown in Figure 3.6(b), the number of training examples in each class may have a large influence on the predictor model. This is termed as an imbalanced classification problem in machine learning. For instance, for the given problem, the number of training examples in class 6 is 998, which is 6.4 % of the number of training examples in class 1. To avoid imbalanced datasets, the method of oversampling is adopted, which duplicates random instances of the minority class to balance the number of data in each class. For the given problem, when the number of examples in a class is less than 5% of the total number of examples in the training set, the class is defined as the minority class and the examples in the minor class are randomly selected to duplicate the new examples for the class.

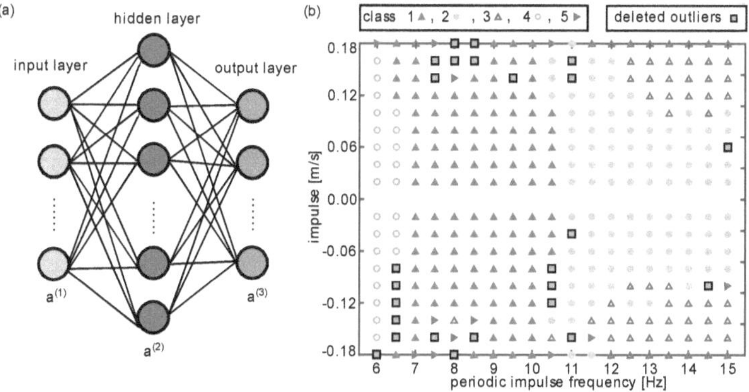

Figure 3.6 (a) Neural network with 3 layers. (b) Classification results for damping constant 0.01 N.s/m and magnet gap 20.55 mm from simulation as functions of the periodic impulse frequency and impulse magnitude in two directions. Class 1 represents intrawell vibration with dominant frequency ratio two, class 2 represents intrawell vibration with dominant frequency ratio one, class 3 represents snap-through vibration with dominant frequency ratio one, class 4 represents intrawell vibration with dominant frequency ratio three, and class 5 represents chaotic vibration. Deleted outliers belongs to other classes with small occurrence possibility

Mathematically, training a neural network is to determine the weights $\Theta$ associated with the minimum value of the cost function. For classification problems, the cost function defined in Eq. (3.15) is utilized to assess the difference of the k-th output unit for the input features of the i-th training example. Once trained, providing new examples $x^i$ to the functions in Eq. (3.16) leads to prediction of the output classifier $\mathbf{a}^{(3)}$, whereby this data-driven model is referred to as a "predictor model".

$$\text{Cost}\left(\left(h_\Theta\left(\mathbf{x}^i\right)\right)_k, y_k^i\right) = \begin{cases} -\log\left(h_\Theta\left(\mathbf{x}^i\right)\right)_k, & \text{if } y_k^i = 1 \\ -\log\left(1 - h_\Theta\left(\mathbf{x}^i\right)\right)_k, & \text{if } y_k^i = 0 \end{cases} \tag{3.15}$$

$$\mathbf{a}^{(1)} = \mathbf{x}^i;\ \mathbf{b}^{(1)} = \begin{bmatrix} 1 & \mathbf{a}^{(1)} \end{bmatrix};\ \mathbf{z}^{(2)} = \Theta^{(1)}\mathbf{b}^{(1)};\ \mathbf{a}^{(2)} = g\left(\mathbf{z}^{(2)}\right);\ \mathbf{b}^{(2)} = \begin{bmatrix} 1 & \mathbf{a}^{(2)} \end{bmatrix};$$

$$\mathbf{z}^{(3)} = \Theta^{(2)}\mathbf{b}^{(2)};\ \mathbf{a}^{(3)} = g\left(\mathbf{z}^{(3)}\right);\ h_\Theta\left(\mathbf{x}^i\right) = \mathbf{a}^{(3)};\ g(z) = \frac{1}{1+e^{-z}}; \tag{3.16a-i}$$

where $x^i$ indicates the input features of the i-th training example; $h_\Theta\left(\mathbf{x}^i\right)$ estimates the probability that $x^i$ belongs to each class and is determined by the forward propagation as in Eqs. (3.16 a-d); a constant 1 is added in the forward propagation to account for the bias existing in the system; $\left(h_\Theta\left(\mathbf{x}^i\right)\right)_k$ is the predicted probability that the input $x^i$ is associated with class $k$; $y_k^i$ is the labeled value to demonstrate the class that the input $x^i$ is truly in, which is 1 if the input features $x^i$ is in class $k$, otherwise $y_k^i = 0$; $\Theta^{(1)}$ and $\Theta^{(2)}$ are weights to be determined through the training process, which are randomly initialized at the start as matrices with the size of $h \times n$ and $m \times (h+1)$; function $g$ is employed to satisfy the (0, 1) limits of the classifier $y_k^i$. In the study, the sigmoid function in Eq. (3.16e) is utilized to map inputs to outputs in the range 0 to 1.

With the predicted probability $h_\Theta(\mathbf{x}^i)$, the class with the maximum probability is the predicted class for the i-th training example by the neural network. For the training example $\mathbf{x}^i$ labeled as in class $k$, thus $y_k^i = 1$. If $\left(h_\Theta(\mathbf{x}^i)\right)_k$ predicts a high possibility around 1 for class $k$, the cost function in Eq. (3.14) is corresponding to a small value around zero. Otherwise, a large value is assigned as a penalty to the cost function. As such, the differences between the labeled and predicted value are measured.

The cost function in Eq. (3.15) can be consolidated to a more compact form as in Eq. (3.17).

$$J_{ik}(\Theta) = -\left[ y_k^i \log\left(h_\Theta(\mathbf{x}^i)\right)_k + \left(1 - y_k^i\right)\log\left(1 - h_\Theta(\mathbf{x}^i)\right)_k \right] \tag{3.17}$$

Further, the total cost function for all training examples and all output units for training the neural network is defined in Eq. (3.18),

$$J(\Theta) = -\frac{1}{N_t}\left[ \sum_{i=1}^{N_t}\sum_{k=1}^{K} J_{ik}(\Theta) \right] \tag{3.18}$$

where $N_t$ is the total number of training examples; $K$ is the total number of output units, which is associated with the total number of classes included in the problem.

An unconstrained minimization function is used to find the minimum of Eq. (3.18) by tailoring the weight matrices. The inputs for the minimization function are the cost function defined in Eq. (3.18) and the random initialization of matrices $\Theta^{(1)}$ and $\Theta^{(2)}$. The corresponding outputs are the trained matrices $\Theta^{(1)}$ and $\Theta^{(2)}$ that are associated with minimum cost function value. With the trained matrices $\Theta^{(1)}$ and $\Theta^{(2)}$, the forward propagation in Eqs. (3.16 a-d) can be employed to predict the class for new input features.

### 3.4.1.2 Assessment of the effectiveness of the physics-guided data-driven model

In order to assess the effectiveness of the mapping formulated by the neural network, the accuracy and F1 score are introduced to quantitively measure the predictor model effectiveness. Eq. (3.19) shows the confusion matrix for the multi-class classification problem. Element  indicates the number of examples in class  that are incorrectly identified as class  when . The diagonal elements demonstrate the number of accurate predictions. With the information provided from the confusion matrix, the accuracy, precision, and recall are defined in Eqs. (3.20 a-c). The accuracy is a measure of correctly identified cases, which is the ratio of the correctly predicted cases to the total number of validation cases. Precision and recall are the ratios between the number of correctly identified instances in a class against the total number of predicted instances and against the total number of the actual cases in the class. For a multi-classification problem, the averages among all classes defined in Eqs. (3.20b,c) are taken as the overall precision and recall for the problem. The F1 score identified in Eq. (3.20d) is the weighted average of the precision and recall to better measure the incorrectly classified cases for imbalanced classification problems.

$$
confusion\,matrix =
\begin{bmatrix}
M_{11} & M_{12} & M_{13} & M_{14} & M_{15} & M_{16} \\
M_{21} & M_{22} & M_{23} & M_{24} & M_{25} & M_{26} \\
M_{31} & M_{32} & M_{33} & M_{34} & M_{35} & M_{36} \\
M_{41} & M_{42} & M_{43} & M_{44} & M_{45} & M_{46} \\
M_{51} & M_{52} & M_{53} & M_{54} & M_{55} & M_{56} \\
M_{61} & M_{62} & M_{63} & M_{64} & M_{65} & M_{66}
\end{bmatrix}
\tag{3.19}
$$

$$
accuracy = \frac{\sum_{m=1}^{6} M_{mm}}{\sum_{i=1}^{6}\sum_{j=1}^{6} M_{ij}}; \quad
precision = \frac{1}{6}\sum_{m=1}^{6}\frac{M_{mm}}{\sum_{i=1}^{6} M_{im}}; \quad
recall = \frac{1}{6}\sum_{m=1}^{6}\frac{M_{mm}}{\sum_{j=1}^{6} M_{mj}}; \quad
F_{1} = \frac{2\cdot recall\cdot precision}{recall + precision}.
$$

$$\tag{3.20a-d}$$

Figure 3.7(a) assesses the numbers of accurate and inaccurate predictions for each class. The total number of examples used in the cross-validation is 20,497. The accuracy and F1 score for the neural network model in predictions are respectively 91.39% and 83.09%. It can be seen from Figure 3.7(a) that the prediction accuracy in class 5 is the lowest. This is because class 5 corresponds with chaotic vibration that occurs without clear delineation among other parametric combinations, such as those class 5 data points shown in Figure 3.6(b). Figure 3.7(b) shows the classification results predicted by the data-driven predictor model by changing the periodic impulse frequency and impulse magnitude for the magnet gap 20.55 mm and damping 0.01 N·s/m. Comparing with Figure 3.6(b), the surrogate model accurately reproduces the demarcation among dynamic regimes and estimates the dominant frequency ratio at the given input features. In addition, the surrogate model correctly captures the influences of the impulse direction on generating snap-through vibrations. As shown in Figure 3.7(b), the periodic impulse in the negative direction realizes snap-through vibrations at a lower periodic impulse frequency around 12 Hz.

From the investigations above, the surrogate model can effectively characterize the dynamics of the nonlinear energy harvesting system. Yet, the criterion and the influence of each design parameter in generating intrawell or snap-through vibration are still implicit in the neural network model. Therefore, in the following section, more insights on selecting design parameters are derived from the trained neural network to assist in determining the optimal strategies for deploying the nonlinear energy harvesting system.

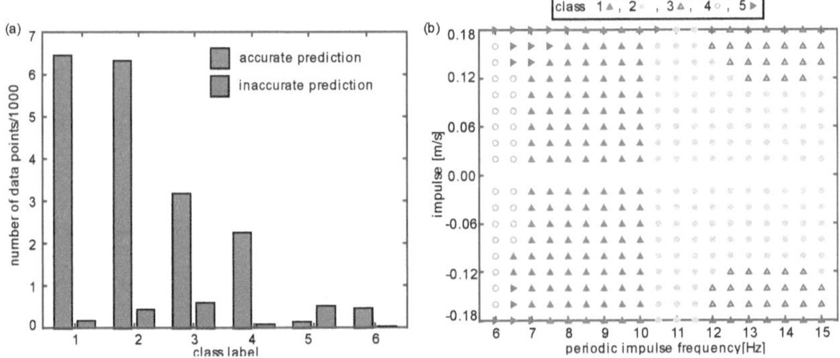


Figure 3.7 (a) Assessment of class prediction accuracy by the supervised predictor model. (b) Classification results for damping constant 0.01 N.s/m and magnet gap 20.55 mm as functions of the periodic impulse frequency and magnitude applied in two directions. Class 1 represents intrawell vibration with dominant frequency ratio two, class 2 represents intrawell vibration with dominant frequency ratio one, class 3 represents snap-through vibration with dominant frequency ratio one, class 4 represents intrawell vibration with dominant frequency ratio three, and class 5 represents chaotic vibration.

## 3.4.2 Optimal strategies from the physics-guided data-driven model for nonlinear energy harvesting systems

To examine the optimal strategies for maximizing DC power delivery, in this section, the design criteria for realizing the snap-through regime are explicitly described based on the surrogate model. For the sake of robustness of harvester designs, sensitivity is then proposed to quantitively measure the influence of the input design parameters on the output dynamics. Moreover, since the properties of the harvesting circuit affect the harvesting energy efficiency, an impedance-based method is also established based on the predicted classification results to estimate the optimal working load for the harvesting circuit.

### *3.4.2*.1 Design criteria for maximum performance and robustness of nonlinear energy harvesters

For the nonlinear energy harvester design, snap-through vibrations are preferred for the high energy output for delivery to a storage battery or electrical load. As shown in Figure 3.6(b), nonlinear energy harvesters require proper parameter combinations to trigger and sustain large amplitude snap-through vibration. An improper design may inhibit continuous snap-through vibration and result in small amplitude intrawell vibrations that do not meet the power requirement for electrical loads. Since the surrogate model predicts the classes associated with the vibration type, in this section the critical values of the design parameters are identified through the decision boundary to help efficiently examine the mechanisms that result in high amplitude snap-through vibrations. In addition, a sensitivity analysis is conducted to determine the integrity of the response boundaries against design uncertainties and inconsistent excitations.

As explained in Section 3.4.1.1, the output of the neural network estimates the possibility of the input features $\mathbf{x}$ in class $k$. The class associated with the highest possibility is the predicted class label for the input features $\mathbf{x}$. As such, as long as the predicted possibility for class $k$ is higher than 0.5, the input features $\mathbf{x}$ are classified in class $k$. Since the sigmoid function in Figure 3.8(a) is utilized as the activation function, the predicted possibility of 0.5 corresponds to an input of zero. Therefore, the input of zero is taken as the boundary among dynamic regimes [97]. Based on the forward propagation in Eqs. (3.16

a-d), Eq. (3.21) is employed to determine the critical values of the design parameters for each dynamic regime.

$$\boldsymbol{\Theta}^{(2)}\mathbf{b}^{(2)} = 0 \tag{3.21}$$

For the given problem, there are a total of 6 classes. Class 3 corresponds to the snap-through vibration with dominant frequency ratio one. Therefore, the third equation derived from Eq. (3.21) corresponds to the critical design parameters to generate snap-through vibrations. Figure 3.8(b) demonstrates the predicted classification results for damping constant 0.01 N.s/m and impulse -0.1 m/s as functions of the periodic impulse frequency and magnet gap. The nonlinear boundary between the snap-through and intrawell dynamic regimes is accurately characterized by the nonlinear equation in Eq. (3.21). As such, the critical need in creating effective tools for bistable harvester design has been addressed. With the criteria determined by Eq. (3.21), the proper design regarding required excitation and system characteristics is identified for maximizing the harvester performance by demarcating sustainable high energy oscillation. Comparing with simulation individually, the established criteria in Eq. (3.21) offer a significant improvement for system designs in terms of conciseness and efficiency. In the following investigations, the mechanism for the transfer of dynamics is further discussed with sensitivity analysis.

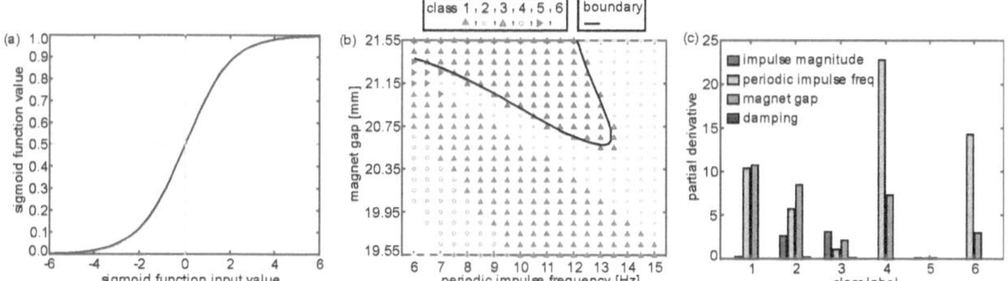


Figure 3.8 (a) Plot of sigmoid function. (b) Classification results for damping constant 0.01 N.s/m and impulse -0.1 m/s as functions of the periodic impulse frequency and magnet gap. Class 1 represents intrawell vibration with dominant frequency ratio two, class 2 represents intrawell vibration with dominant frequency ratio one, class 3 represents snap-through vibration with dominant frequency ratio one, class 4 represents intrawell vibration with dominant frequency ratio three, class 5 represents chaotic vibration, class 6 corresponds to intrawell vibration with dominant frequency ratio four. (c) Partial derivative of each class regrading to input features.

In addition, for the sake of robust designs, the sensitivity analysis is employed to examine the influential design parameters for each dynamic regime. In the following investigations, the partial derivative is defined as sensitivity to measure the influences of the input parameter on the output classes [98]. To determine the partial derivatives, a small perturbation $\varepsilon$ is introduced to one input feature $x_j$ to construct a new feature $\tilde{x}_j$. Here, the input feature $x_j$ is a normalized value representing the j-th feature of $x$ in Eq. (3.14), and $\tilde{x}_j = x_j + \varepsilon$. The new input feature $\tilde{x}_j$ combining with the rest unchanged features constructs a new feature vector $\tilde{x}$. The change on output induced by the perturbation $\varepsilon$ to $x_j^i$ is termed as $\Delta_k$, and $\Delta_k = (h_\Theta(\tilde{x}))_k - (h_\Theta(x))_k$. Therefore, the partial derivative of output $(h_\Theta(x))_k$ in regard to the input feature $x_j$ is determined by the Eq. (3.22a). For $N$ validation examples, the average of partial derivatives defined in Eq. (3.22b) is taken as the sensitivity of the output classes regarding to the input feature.

$$\hat{c}_{kj} = \frac{\partial(h_\Theta(x))_k}{\partial x_j} = \frac{\Delta_k}{\varepsilon}; \quad c_{kj} = \frac{1}{N}\sum_{i=1}^{N}\left|\hat{c}_{kj}^i\right| \tag{3.22a-b}$$

As such, the most important or influential input parameter for each output may have the highest $c$ value.

Figure 3.8(c) presents the sensitivity results for the classes associated with the four input features. According to Figure 3.8(c), the damping constant has the smallest influence on the classification results. For the classes corresponding to intrawell vibrations, namely classes 1, 2, 4, and 6, the periodic impulse frequency and the magnet gap are the two influential factors. As shown in Figure 3.8(b), the vibrations associated with class 4 and 6 can only be generated in a narrower frequency range. This indicates it is more important to select the periodic impulse frequency to ensure the generation of the vibrations in classes 4 and 6. In comparison, the intrawell vibrations in class 1 can be achieved in almost the whole range of the periodic impulse frequency and magnet gap. The sensitivities that measure the influence of the periodic impulse frequency and the magnet gap for class 1 are around a similar scale in Figure 3.8(c).

Moreover, since the generation of snap-through vibrations requires enough energy to overcome the potential energy barrier, the impulse magnitude is the most influential factor for the snap-through vibration represented by class 3 in Figure 3.8(c). If the impulse magnitude is not large enough, the systems may result in intrawell vibrations in class 2. Therefore, the impulse magnitude is a more essential factor in generating intrawell vibration in class 2 comparing with other intrawell classes. In addition, because of the non-resonant nature of the snap-through vibration, the periodic impulse frequency has minor effect on the generation of the snap-through vibration comparing with intrawell vibrations. In terms of the chaotic vibration in class 5, it may be randomly located without evident influential tendency regarding to the design parameters given the results shown in Figures 3.6(b) and 3.7(b). Thus, the chaotic vibration is not sensitive to the changes of input parameters as shown in Figure 3.8(c).

Based on the investigations above, the surrogate model established in Section 3.4.1 is an effective and efficient approach in determining the critical values and sensitivity of design parameters in realizing dynamic regimes. As such, the proper combination of design parameters can be identified to trigger and sustain large amplitude snap-through vibrations.

### *3.4.2.2 An impedance-based method for optimal DC power delivery*

For energy harvesting systems, the impedance matching, or resistance matching theory is usually employed to ensure the optimal power delivery. Studies have concluded the response frequency is more influential in identifying the optimal working load when the energy harvesting system is under harmonic excitation [50] [26]. According to the discussions in Section 3.2.2, the periodic impulse induced responses present similarities with the responses under harmonic excitation. Therefore, in this section, an impedance-based approach is proposed to utilize the dominant frequency component predicted by the surrogate model to help determine optimal loads for all combinations of design parameters.

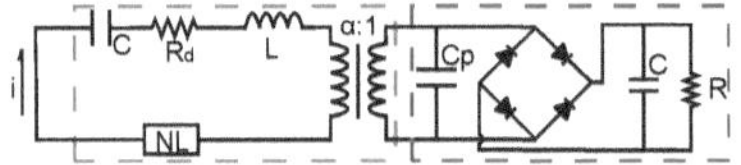

Figure 3.9 Equivalent circuit of a nonlinear energy harvesting system with impulse excitation.


Since the responses under periodic impulse excitation present a single-frequency dominant characterization as discussed in Section 3.2.2, the mechanical-electrical analogy can be extended to energy harvesting systems under periodic impulse excitations. The dominant frequency in responses is provided by the classification results from the surrogate model. Figure 3.9 shows the equivalent circuit of the nonlinear energy harvesting system in Figure 3.1(a). In Figure 3.9, the transformer turn ratio relates to the electromechanical coupling constant $\alpha$. The inductance $L$, resistance $R_d$, and capacitance $C$ in the circuit respectively relate to the mass $L = m$, viscous damping $R_d = c$, and compliance $C_{s1} = 1/k_1$. A general impedance $NL$ block in Figure 3.9 is to characterize the nonlinear effect induced by the nonlinear stiffness $k_3$ and magnetic force $F_{21}$. The study [50] found that the $NL$ block is analogous to a capacitor that has a response-dependent capacitance value. The current through the circuit before the transformer corresponds to the beam tip velocity via $i = \alpha \dot{x}_L$. Because of the ideal impulse assumption, the periodic impulse excitation can be represented by the current $i$ through the circuit, which may experience a sudden change in value at the moment that the impulse is applied.

Since the responses of the nonlinear system under periodic impulse excitation generally can be represented by a single frequency as discussed in Section 3.2, the source and load impedance can be written as Eq. (3.23) with the dominant frequency component estimated from the surrogate model in Section 3.4.1. Here the electrical components inside the red-dashed rectangular in Figure 3.4 are termed as source impedance, correspondingly, the harvesting circuit and the beam internal capacitor in the blue-dashed rectangular are together taken as load impedance.

$$R_L = \frac{A}{\alpha\,\omega_d} \; ; \; X_L = \frac{-B}{\alpha\,\omega_d} \; ; \; R_s = \frac{c}{\alpha^2} \; ; \; X_s = f\left(x_L, \omega_d\right); \tag{3.23a-d}$$

where $R_L, X_L, R_s, X_s$, are respectively the load resistance, load reactance, source resistance, and source reactance; $\omega_d$ is the dominant frequency component in the response, which is predicted by the established surrogate model in Section 3.4.1; A and B identify the characteristic parameters of the SEH circuit, which are defined in Eq. (3.24). Because of the existence of the nonlinear block $NL$, the source reactance depends on the amplitude $x_L$ and frequency $\omega_d$ of the response, which is represented by a general function $f$. A more detailed analysis of the load and source impedance can be found in reference [50].

$$A = \frac{\alpha \sin^2 \Theta}{C_p \pi} \; ; \; B = \frac{\alpha\left(2\Theta - \sin 2\Theta\right)}{2\pi C_p} \; ; \; \Theta = \arccos\left(\frac{\pi - 2\omega_d R C_p}{\pi + 2\omega_d R C_p}\right); \tag{3.24a-c}$$

The study [50] revealed that when $R_S > 2R_L$, the resistance becomes a more influential factor towards the optimal conditions of the harvesting circuit than the reactance. The condition $R_S > 2R_L$ is usually the case given the fact that the electromechanical coupling $\alpha$ is a small value. Therefore, with the dominant frequency component $\omega_d$, the approach to match the load resistance with the source resistance is employed to optimize the delivered power. According to the definition of resistance in Eq. (3.23), the source resistance is determined by the electromechanical coupling and viscous damping. Therefore, the optimal resistance $R$ can be determined by minimizing the difference between $R_S$ and $R_L$

$$R_o = \arg\min_{R}\left(|R_L - R_S|\right) \tag{3.25}$$

Figure 3.10 (a) shows the optimal resistance for the nonlinear harvesting circuit as a function of the periodic impulse frequency for impulse -0.058 m/s, magnet gap 21 mm, and damping constant 0.01 N.s/m. The red circle data labeled simulation are calculated by sweeping through a range of resistances with simulations. Instead, the green triangles named prediction are from the resistance matching approach based on Eq. (3.25). As shown in Figure 3.2(c), when the periodic impulse frequency is around 8 Hz, the numerical response is indeed characterized by multiple frequencies. Yet, based on the definition of the dominant frequency in Section 3.2.2, the classification results from the surrogate model only considers the influence of the single frequency component. Since the optimal resistance depends on the frequencies of response, there is a disagreement on the predicted optimal resistance around 8 Hz.

In addition, depending on the system parameters and initial conditions, there is a coexistence of snap-through and intrawell vibrations under certain excitation conditions for a nonlinear system. For example, as shown in Figure 3.10(b), when the periodic impulse frequency is 10.75 Hz, both snap-through and intrawell vibrations can be introduced by changing the resistance. In this case, the optimal resistance corresponds to the one that can introduce snap-through vibration [50]. Since the surrogate model in Section 3.4.1 assumes the resistance as a constant to characterize the dynamic regimes of the nonlinear energy harvesting system, the transition from intrawell vibration to snap-through vibration by changing resistance is not included. The impedance analysis based on the classification results from the surrogate model may not be consistent with the values from simulations that consider the transition of vibration types, such as cases at frequencies 10.75 Hz and 12 Hz in Figure 3.10(a).

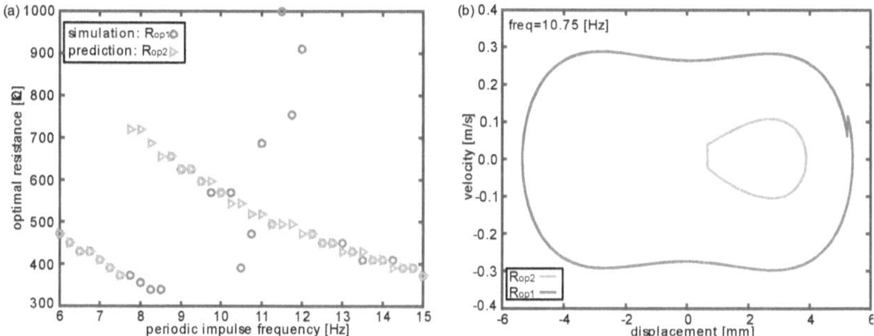


Figure 3.10 (a) Optimal resistances across excitation frequencies. (b) Phase portraits with two optimal resistances at periodic impulse frequency 10.75 Hz.

Overall, the prediction results give good approximations of the optimal resistances. Except for the small group of resistances that are associated with co-existent intrawell and snap-through oscillations, the maximum relative error of 'prediction' comparing with simulation is less than 5%. It is reasonable to employ the simple resistance matching approach to approximate optimal resistances for the harvesting circuit.

85

With the emerging implementation of vibration energy harvesting systems to support the Internet-of-Things devices, significant operational data may be collected in the field. A synthesis of first-principles models and machine learning algorithms can provide real-time optimization of the energy harvesting platforms to ensure sustainable operation of power delivery. This research demonstrates one example of such synthesis, although it is anticipated considerable progress may still be made in this emerging field.

**3.5 Conclusion**

This research integrates data-driven analysis with the physical insights of a first-principles model to examine strategies to ensure nonlinear energy harvesting systems that are subjected to periodic impulse excitation deliver peak electrical power to loads. The investigations indicate that the direction of the periodic impulse plays a significant factor for electrical power generation, due to the asymmetry of system energy associated with the nonlinearities. A predictor model is created through a neural network analysis of simulation data to identify design and excitation parameter regimes that ensure high amplitude snap-through dynamics are induced, providing best DC current delivery. The accuracy of the predictions are confirmed through experiments and against cross-validation simulation data. The demarcation of parameters that result in snap-through are then probed through a first-principles model to confirm that optimal circuit design conditions, specifically the resistance, suggested by the predictor model agree with the principle of impedance matching. This research demonstrates a successful synthesis of data-driven and first principles models to guide attention to optimal designs of nonlinear energy harvesting systems subjected to periodic impulse excitation.

## Chapter 4 Electrical power management and optimization with nonlinear energy harvesting structures

In recent years, great advances in understanding the opportunities for nonlinear vibration energy harvesting systems have been achieved attention to either the structural or electrical sub-systems. Yet, a notable disconnect appears in the knowledge on optimal means to integrate nonlinear energy harvesting structures with effective nonlinear rectifying and power management circuits for practical applications. Motivated to fill this knowledge gap, this research employs impedance principles to investigate power optimization strategies for a nonlinear vibration energy harvester interfaced with a bridge rectifier and buck-boost converter. The frequency and amplitude dependence of the internal impedance of the harvester structure challenge conventional impedance matching concepts. Instead, a system-level optimization strategy is established and validated through simulations and experiments. Through careful studies, the means to optimize the electrical power with partial information of the electrical load is revealed and verified in comparison to the full analysis. These results suggest that future study and implementation of optimal nonlinear energy harvesting systems may find effective guidance through power flow concepts built on linear theories despite the presence of nonlinearities in structures and circuits.

## 4.1 Introduction

The possibilities for wireless, self-sufficient electrical supplies to advance myriad applications in an emerging "internet of things" are recently well established [99] [100]. For instance, self-sufficient power resources would propel new and future practices of structural health and condition monitoring, wild life tracking, and precision agriculture [99] [101] [102]. In addition, achievements in energy-efficient circuit design [103] [104] suggests the possibilities for a transformative "energy harvesting" infrastructure are closer at hand than ever.

Among all kinds of ambient energy sources, vibrational energy is a desirable choice to harvest for the abundance and persistence of kinetic energy in applications. Because many energy harvesting devices are composed from structural oscillator designs, the narrow frequency of resonance for linear harvester systems is unfavorable for sustainability in electrical power delivery. As a result, a wide variety of methods have been introduced to increase the frequency range of effective energy capture in vibration energy harvesters. For example, self-tuning [105], harvester arrays [72], mechanical stoppers [6], nonlinearity [106] [107] [8] and other concepts have been investigated. Bistable nonlinearities have attracted broad attention due to the non-resonant nature of oscillation that is less susceptible to performance deterioration in low-level vibration environment [42] [15] [12] [108]. A bistable oscillator has two statically-stable equilibria, which makes it possible to exhibit three classes of vibration in consequence to single frequency excitation: snap-through, quasi-linear, and chaotic vibration [8]. The non-resonant nature of snap-through enables it to be triggered in a diverse range of excitation environments, especially at low frequencies characteristic of practical ambient vibrations [9]. In addition, snap-through may yield large velocities of harvester motion that corresponds to high output power. For example, Erturk et al. (2009) demonstrated a 200% increase in the open-circuit voltage under harmonic excitation with snap-through vibrations of a bistable energy harvester, while Ferrari et al. (2010) found an 80% increase in the RMS voltage with white-noise excitation.

Since microelectronic devices require a regulated DC input voltage, a rectifying circuit is necessary for energy harvesters to convert the oscillating AC signals to DC power. There are two classes of rectification: passive and active rectification [109]. Comparing with passive methods, active methods may improve the power conversion efficiency at low power levels [110]. On the other hand, active rectification may require careful design to overcome challenges such as "cold start" [111]. Therefore, despite the possible improvements from active rectification methods, passive rectification such as full-wave diode bridges are widely employed for rectification needs in vibration energy harvesting investigations. To improve power capture, such rectifiers may be modified via switching strategies. For instance, the parallel-SSHI (synchronized switching harvesting on inductor) presented by Guyomar et al. (2005) adds an inductor and a switch in parallel with the linear piezoelectric element electrodes to the input of the rectifier bridge so as to enhance power potentially by 900% according to the switch-controlled current flow [112]. Other types of nonlinear circuit like series-SSHI and optimized series-SSHI may improve the output power as well [113] [114] [115].

Yet for a selected circuit, the optimal output power is only achieved when the load impedance matches the source impedance. Liang and Liao (2012) characterized such influences for a linear piezoelectric energy harvester to reveal optimal working condition for a variety of the nonlinear rectification circuits [26]. On the other hand, rectified voltages from piezoelectric energy harvesters may significantly exceed the required operational voltages of rechargeable batteries and microelectronics. As such, a DC-DC converter is introduced after the piezoelectric voltage rectification stage to provide the needed power management [116]. Lefeuvre et al. (2007) presents the roles of a buck-boost converter working in the discontinuous current mode (DCM), revealing that the nonlinear converter appears as a resistive load impedance with respect to the linear piezoelectric energy harvester platform regardless of the real load resistance bearing the DC power. These findings demonstrate that the nonlinear rectification and power management are crucial to understand in order to effectively implement linear piezoelectric energy harvesters in the conventional resonant vibration mode.

Considering the state-of-the-art, nonlinearities in the vibration energy harvesting structure or nonlinearities in the rectification and power management stages may be leveraged for effective DC power delivery. On the other hand, understanding of the suitable integration of these sub-systems is lacking so that a disconnect of knowledge exists on the system-level harvester implementation. In order to bridge the gap in the understanding of optimal use of nonlinear energy harvesters with nonlinear rectification and power management circuits, this research takes a first step forward to study the structural-electrical interactions manifest in a nonlinear energy harvester coupled with a passive diode bridge rectifier and a buck-boost DC-DC converter. By harnessing principles of impedance, new insights are revealed on practical strategies for system optimization.

This report is organized as follows. The next section studies the internal impedance of the nonlinear harvester structural platform. Then, rectification and buck-boost converter stages are taken into detailed consideration. Following validation efforts, the impedance changes observed at the system-level are revealed and the optimal working conditions for a buck-boost DC-DC converter integrated with the nonlinear harvester are uncovered through a rigorous impedance-based analysis. Finally, a summary of key insights from this work concludes the report.

## 4.2 Source/load modeling of nonlinear energy harvesting system

### 4.2.1 Energy harvester platform

The structural platform of the experimental energy harvesting system considered in this work is shown in Figure 4.1(a). A piezoelectric cantilever (PPA-2014, Mide Technology) has the clamped end affixed to an electrodynamic shaker (APS Dynamics 400). The shaker is driven by a controller (Vibration Research Controller VR9500) and amplifier (Crown XLS 2500). Laser displacement sensors (Micro-Epsilon ILD-1420) are used to measure the absolute displacements of the beam tip and shaker table. An accelerometer (PCB Piezotronics 333B40) is used to provide feedback for the shaker controller. The nonlinearity exploited in this research is a bistability realized by magnetoelastic effects. Specifically, at the free end of the piezoelectric cantilever a steel extension is added to decrease the lowest order natural frequency of the beam and introduce nonlinearity by magnetic forces from an adjacent pair of neodymium magnets. The total mass of the extension is 9 g. The length $L$ of the cantilever beam and steel extension is 60.8 mm. With careful adjustment of the distances $\delta$ and $\Delta$ shown in the Figure 4.1(b), the bistable nonlinearity is realized [117] [118] [41] [119]. Based on the survey of state-of-the-art developments in Section 4.1, the advancements provided by a bistable vibration energy harvester are associated with the snap-through behavior. Therefore, in the following investigations the study places attention on scrutinizing the DC power delivery from the snap-through response.

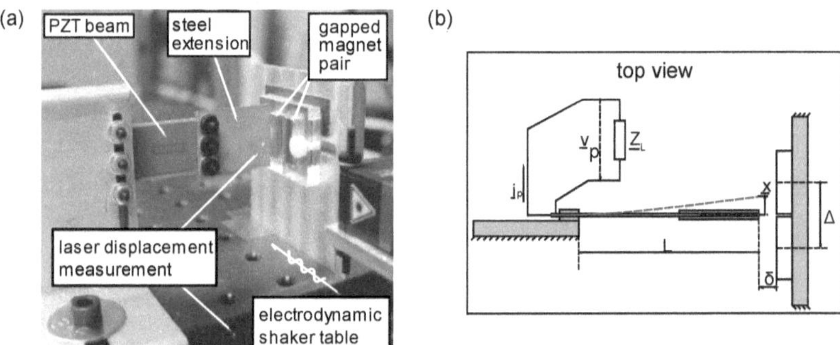


Figure 4.1 (a) Photograph and (b) schematic of experimental setup.

### 4.2.2 Governing equations of motion for the energy harvesting system

In this work, the frequencies of base acceleration are around the lowest order linear natural frequency of the piezoelectric cantilever, so that the deflection of the cantilever is primarily in the first vibration mode [2]. Therefore, a single degree-of-freedom model is adopted to study the nonlinear energy harvesting system. The governing equations are expressed as

$$m\ddot{\underline{x}} + d\dot{\underline{x}} + k_1\left(1-p\right)\underline{x} + k_3\underline{x}^3 + \alpha\underline{v}_p = -m\ddot{\underline{z}} \tag{4.1a}$$

$$C_p\dot{\underline{v}}_p + \underline{i}_l = \alpha\dot{\underline{x}} \tag{4.1b}$$

In Eq. (4.1), $\underline{x}$ is the beam tip displacement relative to the motion of the base displacement $\underline{z}$; $m$, $d$, $k_1$, and $k_3$ are the equivalent lumped mass, viscous damping, linear stiffness, and nonlinear stiffness corresponding to the first vibration mode; $p$ is the load parameter, which is used to indicate the influence of magnetic forces on reducing the linear stiffness; $\alpha$ is the electromechanical coupling constant; $c_p$ is the internal capacitance of the piezoelectric beam; $\underline{v}_p$ is the voltage across the piezoelectric beam electrodes; $\underline{i}_p$ is the corresponding current that passes into the harvesting circuit impedance $\underline{Z}_L$ as exemplified in Figure 4.1(b). The overdot operator indicates differentiation with respect to time $t$.

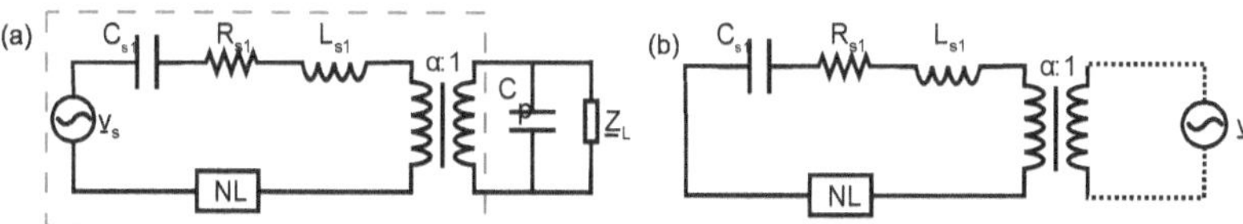


Figure 4.2 (a) Equivalent circuit of a base-excited nonlinear energy harvesting system and (b) equivalent circuit for the internal source impedance determination.

Studies have shown the analogy between mechanical and electrical systems in the energy harvesting literature [26] [120] [121]. Here, such concept is employed to identify an equivalent electrical system shown in Figure 4.2(a) that represents the vibration energy harvesting system. The electromechanical coupling constant $\alpha$ is interpreted as a transformer turn ratio. In Figure 4.2(a), $\underline{v}_{s1}$ corresponds to the base acceleration via $\underline{v}_{s1} = -m\underline{\ddot{z}}$. The inductance $L_{s1}$, resistance $R_{s1}$, and capacitance $C_{s1}$ in the circuit respectively relate to the mass $L_{s1} = m$, viscous damping $R_{s1} = d$, and compliance $C_{s1} = 1/k_1$. For the nonlinearity in the energy harvesting system characterized by the negative linear stiffness $-pk_1$ and nonlinear stiffness $k_3$, a general impedance $NL$ block is shown in Figure 4.2(b) that will be clearly defined in the subsequent modeling of this work. In the following study, the components in the red dashed box are collectively considered to be the source impedance $\underline{Z}_s$ of the nonlinear energy harvesting system. The current through this source impedance is related to the relative velocity of the beam tip via $\underline{i} = \alpha \underline{\dot{x}}$.

### 4.2.3 Experimental system identification

To identify the parameters of the nonlinear cantilever, an impulsive ring-down test is first conducted to ensure the symmetry of the two static equilibria via identical linear natural frequencies of oscillation around each stable equilibrium. The equivalent lumped mass $m$ and the linear stiffness $k_1$ are calculated by classical relations [67]. After determining the natural frequency $\omega_0$ from the ring down experiments and measuring the static equilibria $x^*$, the load parameter $p$ and nonlinear stiffness $k_3$ are found by

$$p = 1 + \omega_0^2 m / 2k_1 \; ; \; k_3 = k_1 (p-1)/(x^*)^2 \qquad (4.2a,b)$$

The identified parameters of the nonlinear energy harvester are shown in Table 4.1.

Table 4.1 Experimentally identified system parameters.

| $m$ (g) | $d$ (N s/m) | $k_1$ (N/m) | $p$ (dim) | $k_3$ (MN/m$^3$) | $C_p$ (nF) | $\alpha$ (mN/V) |
|---|---|---|---|---|---|---|
| 9.45 | 0.194 | 722 | 1.27 | 86 | 88 | 1.4 |

### 4.2.4 Source impedance characterization

The original source $\underline{v}_{s1}$ in Figure 4.2(a) is removed to study the source impedance $\underline{Z}_s$ of the harvester itself. Instead, an AC voltage $\underline{v}$ is connected to the output terminal, which is referred to as the driven voltage shown in Figure 4.2(b). Then, the governing equation is

$$m\underline{\ddot{x}} + d\underline{\dot{x}} + k_1\left(1-p\right)\underline{x} + k_3\underline{x}^3 = \alpha\underline{v} \tag{4.3}$$

The driven voltage is

$$\underline{v} = -\underline{V}\cos\left(\underline{\omega}t\right) \tag{4.4}$$

The governing Eq. (4.3) after non-dimensionalization is

$$x'' + \eta x' + \left(1-p\right)x + \beta x^3 = \kappa v \tag{4.5}$$

The corresponding non-dimensional parameters are defined as follows:

$$x = \underline{x}/x_0 \;;\; V = \underline{V}/V_0 \;;\; \tau = \omega_0 t$$

$$\beta = k_3 x_0^2 / k_1 \;;\; \kappa = \alpha V_0 / k_1 x_0$$

$$\omega_0 = \sqrt{k_1/m} \;;\; \omega = \underline{\omega}/\omega_0 \;;\; \eta = d/m\omega_0 \tag{4.6a-h}$$

Here $x_0$ and $V_0$ are characteristic length and voltage, respectively, that are defined to be 1 mm and 1 V. These selections are made to keep the non-dimensional response values around the order of 1.

Using principles of harmonic balance [122], the solution to (5) may be approximated as

$$x\left(\tau\right) = c\left(\tau\right) + a\left(\tau\right)\sin\left(\omega\tau\right) + b\left(\tau\right)\cos\left(\omega\tau\right) \tag{4.7}$$

The coefficients of the constant, sine, and cosine terms are collected after substituting Eq. (4.7) into Eq. (4.5). Since only the fundamental frequency of vibration is considered in (4.7), the higher order harmonics are neglected. In addition, the coefficients are assumed to vary slowly in non-dimensional time. The resulting system of equations collecting together constant and sinusoidal terms is

$$-\eta c' = c\left(1 - p + \frac{3}{2}\beta r^2 + \beta c^2\right)$$

$$-\eta a' + 2\omega b' = a\Lambda - bX \qquad (4.8\text{a-c})$$

$$-2\omega a' - \eta b' = aX + b\Lambda + V\kappa$$

The following terms are identified

$$r^2 = a^2 + b^2; \quad \Lambda = 1 - p - \omega^2 + \frac{3}{4}\beta r^2 + 3\beta c^2; \quad X = \eta\omega \qquad (4.9\text{a-c})$$

The steady-state behavior of the nonlinear equivalent source is determined by first solving Eq. (4.8a) to yield

$$c = 0 \quad \text{or} \quad c^2 = \frac{p-1}{\beta} - \frac{3}{2}r^2 \qquad (4.10\text{a-b})$$

Here, the snap-through behavior is related to $c = 0$ so that the displacements of the piezoelectric cantilever undergo zero-mean oscillations. Therefore, only $c = 0$ is considered in the following evaluations.

Then, combining Eq. (4.8b) and (4.8c), the snap-through response is studied through solutions to the roots of the polynomial (4.11).

$$\left(\Lambda^2 + X^2\right)r^2 = \left(V\kappa\right)^2 \qquad (4.11)$$

Eq. (4.11) is a cubic polynomial in terms of $r^2$. Complex and negative solution are not physically meaningful. Stability of the roots is assessed according to the eigenvalues of the associated Jacobian from Eq. (4.8) [122].

For the circuit shown in Figure 4.2(b), the voltage and current are expressed in a complex exponential notation to characterize the source impedance $\underline{Z}_s$.

$$\underline{Z}_s = \frac{v}{\underline{i}} = \frac{-Ve^{j\omega t}}{\alpha \underline{\dot{x}}} = \frac{-V_0}{\alpha x_0 \omega_0}\frac{Ve^{j\omega \tau}}{\omega r \cdot e^{j(\omega \tau - \varphi_i)}} = \frac{-V_0 Ve^{\varphi_i}}{\alpha x_0 \omega_0 \omega r} \tag{4.12}$$

where

$$\tan \varphi_i = -b/a \tag{4.13}$$

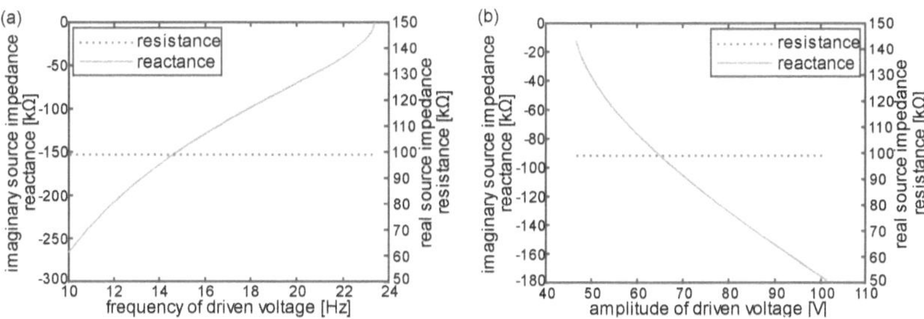

Figure 4.3 (a) Driven voltage frequency and (b) amplitude influences on the source impedance.

Figure 4.3 presents source impedance results the characterize the impedance change for change in the driven voltage. The experimentally identified system parameters from Table 4.1 are used to generate Figure 4.3 using the aforementioned model. The fixed amplitude of the driven voltage for Figure 4.3(a) is 51 V and the fixed driven voltage frequency for Figure 4.3(b) is 22 Hz. Since attention is only on snap-through behavior, the source impedance does not exist in the whole range of driven voltage frequency and amplitude for Figure 4.3. In addition, it is observed that the reactance is frequency- and amplitude-

97

dependent. Such dependencies are evidence of the cubic nonlinearity and presence of negative linear stiffness via the bistable nonlinearity.

In the steady state analysis, only the fundamental frequency vibration is considered. Therefore, a linearized version of the system (5) is

$$x'' + \eta x' + \left(1 - p + \frac{3\beta}{4} r^2\right) x = \kappa v \tag{4.14}$$

From Eq. (4.14), the roles that the nonlinear and negative stiffness terms play to culminate in the system behaviors become clear. The terms modeled as the generic $NL$ impedance in Figure 4.2 are analogous to a capacitor that has a response-dependent capacitance value. This also explains the amplitude-dependent source reactance in Figure 4.3. In contrast, the resistance in Figure 4.3 is independent of the frequency and amplitude of the driven voltage, due to the nature of the viscous damping-based resistance $R_{s1} = d$ in Figure 4.2.

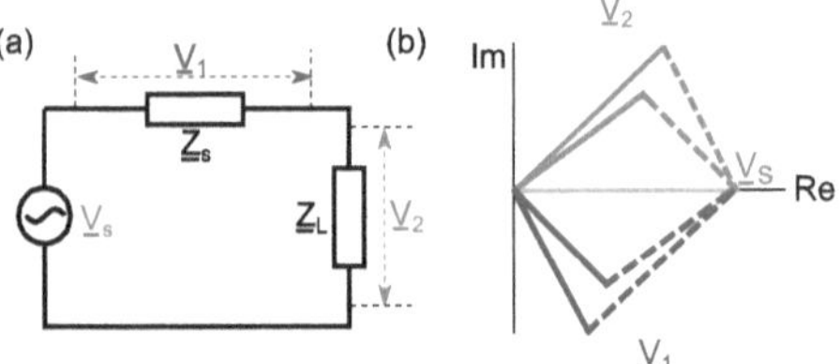


Figure 4.4 (a) The simplified circuit schematic for a nonlinear harvesting system and 4 (b) voltage shown in complex plane.

For any harmonically driven electrical system, optimal power delivery to a load is achieved when the load impedance is the complex conjugate of the source impedance. A suboptimal approach is resistive impedance matching that may be relevant for vibration energy harvesting systems [120]. Figure 4.4(a) shows a simplified circuit schematic for a vibration energy harvesting system. The source voltage amplitude is $V_s$, characteristic of the base acceleration. The corresponding voltages across the source impedance and load impedance are respectively $V_1$ and $V_2$. Figure 4.4(b) presents these voltages in the complex plane. By Kirchhoff's law, the combination of $V_1$ and $V_2$ is equal to $V_s$. For the nonlinear system examined here, if the load impedance is changed to match the source impedance, the voltage across the load impedance is changed due to the amplitude-dependent nature of the system. This correspondingly changes $V_1$. As a result, a standard practice of impedance matching is significantly challenged. Therefore, in this research the investigation of optimal power delivery from the nonlinear energy harvesting system with power management circuitry requires studious consideration of the platform-circuit system in operation, rather than sole attention to one subcomponent.

In the following Section 4.3, a rectifier and buck-boost converter are introduced to the load impedance model of Figure 4.4(a) to examine the system-level behaviors and optimal DC power delivery conditions.

## 4.3 Analytical model formulation and solution for the nonlinear energy harvesting system with DC power management

### 4.3.1 System model and approximate solution to the governing equations

The governing equations of motion (1) are non-dimensionalized to be

$$x'' + \eta x' + (1-p)x + \beta x^3 + \kappa v_p = -z''$$

$$v_p' + i_p = \theta x' \tag{4.15a-b}$$

The non-dimensional base acceleration is

$$-z'' = a\cos\omega\tau \tag{4.16}$$

Here $a$ is the normalized base excitation of $\underline{a}$. Additional non-dimensional parameters are defined

$$\theta = \alpha x_0 / C_p V_0 \,;\; a = \underline{a}m/k_1 x_0 \,;\; I_p = \underline{I}_p / C_p V_0 \,\omega_0 \,;\; V_p = \underline{V}_p / V_0 \,. \tag{4.17a-c}$$

The capital terms $I_p$, $\underline{I}_p$, $V_p$, and $\underline{V}_p$ are the amplitudes of the corresponding variables $i_p$, $\underline{i}_p$, $v_p$, and $\underline{v}_p$.

Assuming that the nonlinear energy harvester responds with oscillatory displacements with the same frequency as the base acceleration, the non-dimensional displacement is given by

$$x(\tau) = k(\tau) + h(\tau)\sin(\omega\tau) + g(\tau)\cos(\omega\tau) \tag{4.18}$$

Prior to having means to approximately solve for the system response, the characteristics of the non-dimensional voltage $v_p$ and current $i$ must be identified.

Figure 4.5 presents the interface circuit to which the nonlinear piezoelectric cantilever is attached. The circuit includes a rectifying bridge, smoothing capacitor $C_1$, and a buck-boost converter. The components in the green dashed box constitute the buck-boost converter that modulates the voltage across the load $R$ so as to permit practical implementation of the DC power. Such power management is required because piezoelectric energy harvesters may generate high rectified voltages (>20 V) whereas many rechargeable batteries or microelectronics use low, standardized voltage levels such as 3 V, 9 V, and so on. The inductor, capacitor, resistor, and diode in the buck-boost converter are labeled $L$, $C_2$, $R$, and $D$ respectively. The n-type MOSFET $M_N$ in the converter acts like a switch.

A pulse width modulation (PWM) signal in red is sent to the gate of the MOSFET to control the open state.

In general, a buck-boost converter may run in two modes: discontinuous current mode (DCM) and continuous current mode (CCM). Compared to CCM, the switching losses are much less for DCM [109], so that attention in this research is solely on the DCM operation. A buck-boost converter running in DCM will realize an impedance analogous to an equivalent resistance [123]

$$R_{eq} = \frac{2 \cdot L \cdot f_s}{D^2} \tag{4.19}$$

where $L$ represents the inductance in the circuit; and the switching frequency $f_s$ and duty cycle $D$ are properties of the PWM signal. Since the behavior of the converter is not influenced by the real load $R$ in Figure 4.5, the optimal working condition for a buck-boost converter will not change for different load $R$. Therefore, the DCM operation is the preferred choice for energy harvesting applications.

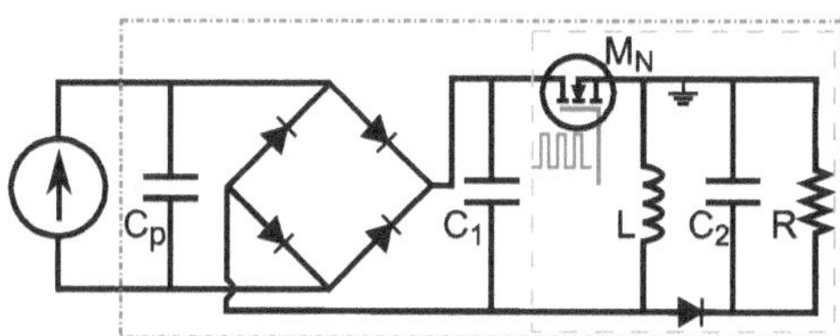


Figure 4.5 The nonlinear vibration energy harvester with a buck-boost converter.

Since the buck-boost converter in DCM can be replaced by an equivalent resistance $R_{eq}$ (19), the interface circuit is reduced to a rectifier and an $R_{eq}C_1$ circuit. Based on previous studies [14] [26], the non-dimensional voltage $v_p$ across the piezoelectric beam capacitance $C_p$ is

$$v_p = \left[ \frac{-g(\tau)}{\pi} \theta \sin^2 \Theta + \frac{h(\tau)}{2\pi}(2\Theta - \sin 2\Theta) \right] \sin \omega\tau$$

$$+ \left[ \frac{h(\tau)}{\pi} \theta \sin^2 \Theta + \frac{g(\tau)\theta}{2\pi}(2\Theta - \sin 2\Theta) \right] \cos \omega\tau \tag{4.20}$$

In Eq. (4.20), only the fundamental harmonic of voltage is considered on the basis that the fundamental contributes a significant proportion of the Fourier series reconstruction of the actual voltage signal [14] [26]. The parameters in Eq. (4.20) are:

$$\Theta = \arccos\left(\frac{\pi\rho - 2\omega}{\pi\rho + 2\omega}\right) ; \quad \rho = \frac{1}{R_{eq}C_p\omega_0} \tag{4.21a-b}$$

Then, Eq. (4.18) and (4.20) are substituted into Eq. (4.15a). Using the method of harmonic balance, the Eq. (4.22) is then obtained. The roots of the polynomial (4.22) are the squared amplitudes of the non-dimensional displacement $n^2$. The voltage $v_p$ is thereafter determined from (4.20).

$$\left(\Lambda_1^2 + X_1^2\right)n^2 = a^2 \tag{4.22}$$

The terms in (4.22) are

$$n^2 = g^2 + h^2 ; \quad \Lambda_1 = 1 - p + B\kappa - \omega^2 + 3\beta n^2/4 + 3\beta k^2 ; \quad X_1 = A\kappa + \eta\omega ; \quad A = \frac{\theta \sin^2 \Theta}{\pi} ; \quad B = \frac{\theta(2\Theta - \sin 2\Theta)}{2\pi}$$

$$\tag{4.23a-e}$$

When a buck-boost converter operates in DCM, the voltage across the resistor $R$ in Figure 4.5 is calculated by [124]

$$V_O = V_p V_0 D \sqrt{\frac{R}{2Lf_s}} \tag{4.24}$$

Then the DC output power is

$$P_O = \frac{V_O^2}{R} = \frac{V_p^2 V_0^2 D^2}{2Lf_s} \tag{4.25}$$

The $V_p$ is the amplitude of $v_p$; $V_O$ is the DC voltage across the resistor $R$, and $P_O$ is the corresponding DC output power.

### 4.3.2 Determination of load impedance

The components in Figure 4.5 in the red dash-dot box are taken to be the load impedance $Z_L$ for the nonlinear energy harvesting system. The $v_p$ is defined by (4.20). By (4.15b), the non-dimensional current that flows into the load is

$$i = \theta x' = \omega\theta \left[ h\cos(\omega\tau) - g\sin(\omega\tau) \right] \tag{4.26}$$

In the following calculation, current and voltage are expressed in the complex exponential form for load impedance calculation.

$$Z_L = \frac{v_p}{i} = \frac{\sqrt{A^2 + B^2}\, e^{j(\varphi_i - \varphi_v)}}{\omega\theta} \tag{4.27}$$

The phase angles are

$$\tan\varphi_v = \frac{-gA + hB}{hA + gB}\;;\quad \tan\varphi_i = -\frac{g}{h} \tag{4.28a-b}$$

The corresponding dimensioned load impedance is thus

$$Z_L = \frac{V_0\theta}{\alpha x_0 \omega_0} Z_L \tag{4.29}$$

Which yields the load resistance $R_L$ and load reactance $X_L$ :

$$R_L = \frac{V_0 \mathrm{A}}{\alpha x_0 \omega_0 \omega} \; ; \; X_L = \frac{-V_0 \mathrm{B}}{\alpha x_0 \omega_0 \omega} \tag{4.30a-b}$$

## 4.4 Validation of the analytical model formulation and solution

### 4.4.1 Power management circuit design

The specific circuit components used for electrical power management are listed in Table 4.2. In the experimental validation, the PWM signal is provided by a function generator (Siglent SDG 1025) instead of a self-powered oscillator [120]. This approach to PWM signal generation inhibits complication of the assessment of the principles studied here regarding analytical prediction of the optimal DC power delivery conditions. Future work will consider methods of self-powering the power management circuit.

Table 4.2 Circuit components used for rectifier and buck-boost converter.

| Component | Value or part |
|---|---|
| Rectifier diodes | 1N4148( $V_{d1} = 1$ V; $R_{d1} = 100\,\Omega$ ) |
| Diode $D$ | Schottky 1N5820G ( $V_{d2} = 0.37$ V; $R_{d2} = 0.37\,\Omega$ ) |
| $C_1(C_2)$ | 10 μF (470 μF) |
| $L$ | 1 mH( $R_L = 0.216\,\Omega$) |
| $M_N$ | IRLZ44N ( $V_{ds} = 1.3$ V; $R_{ds} = 0.022\,\Omega$ ) |
| $R$ | 2     kΩ |

## 4.4.2 Experimental validation and comparison to analytical and numerical results

The experimental system is used to validate the proposed analytical formulation for DC power optimization. In addition, verification of the accuracy of the approximate analytical solution is made by direct consideration of the equations of motion via a MATLAB Simulink model of the equivalent circuit, Figure 4.4(a) where the $\underline{Z}_s$ is given by (4.12) and $\underline{Z}_L$ is given by (4.29). Runge-Kutta numerical integration is employed for any given base acceleration combination of amplitude and frequency. For such combination, 8 randomly distributed sets of initial conditions are prescribed for the currents in order to ensure that the simulations identify all possible steady-states of response in the nonlinear harvester structure and circuit system. The calculation time for each harmonic excitation condition is over 200 periods to ensure the steady-state is reached. The needs to utilize such long simulation times and multiple initial conditions make the simulation significantly more costly to compute than the analysis.

In order to take the losses of the electrical components into account, the respective forward voltage drops and the respective on resistances for diodes and the MOSFET are employed in the analytical model and Simulink model. The values for the parameters are given in Table 4.2.

In analysis, the output voltage with electrical losses can be written as [124],

$$\underline{V}_O = \frac{V_{p1} V_0 D}{D_2} - V_{d2} - \underline{i}_L R_L - \frac{D}{D_2} \underline{i}_L \left( R_L + R_{ds} \right) \tag{4.31}$$

The $D_2$ relates to the time that the current through the inductor drops to zero. $V_{d2}$ is the voltage drop of the diode, $R_L$ and $R_{ds}$ are the on resistance of the inductor and MOSFET, $i_L$ is the current through the inductor, $V_{p1}$ is the input voltage of the buck boost converter, which is the voltage across the capacitor $C_1$. Considering the losses caused by the rectifier bridge, the $V_{p1}$ can be expressed as

$$V_{p1} = V_p - 2V_{d1} - 2i_L R_{d1} \tag{4.32}$$

The $V_{d1}$ and $R_{d1}$ are the voltage drop and on resistance of the diode in rectifier bridge. Since the currents $i_L$, $i_p$ and the on resistances $R_L$, $R_{ds}$ are relatively small values, the Eqs. (4.31) (4.32) are approximated by

$$V_O = \frac{V_{p1} V_0 D}{D_2} - V_{d2} \tag{4.33}$$

$$V_{p1} = V_p - 2V_{d1} \tag{4.34}$$

The expression for $D_2$ adopted is

$$D_2 = \sqrt{\frac{2Lf_s}{R}} \tag{4.35}$$

Figure 4.6(a) presents the amplitude of the harvester displacement determined by analysis and simulation. The base acceleration amplitude is 9.8 m/s² and the excitation frequency is around the linearized natural frequency 30 Hz. Since greater values of the buck-boost converter switching frequency result in greater losses, in this work the switching frequency remains at 1 kHz in all evaluations. As a result of the switching frequency selection, it is recognized that the preferred range of duty cycle may be less than 1% [120]. For the specific result presented in Figure 4.6(a), the duty cycle is 0.1%. Based on the analytical and simulation results, the large amplitude snap-through responses and small amplitude intrawell responses coexist in the lower frequency range, such around 10 to 17 Hz. In the frequency range around 20 Hz, only snap-through responses occur. The simulation predicts a narrower range of frequencies across which only snap-through exists compared to analysis [122]. At higher frequencies such as 30 Hz, both simulation and analysis agree that an intrawell response alone may occur. In the range from 23 to 30 Hz the simulations yield aperiodic oscillations that result in different amplitudes of displacement based on the initial conditions of the simulation. In contrast, the analysis, based on the method of harmonic balance and steady-state assumptions, predict that either an intrawell or snap-through behavior may occur. These trends are seen elsewhere in the literature that compare analytical predictions to numerical simulations of nonlinear dynamics of vibration energy harvesters around frequency bands of coexistent responses [125] [14] .

Figure 4.6(b) shows the corresponding experimental results. Compared to the model results, the large amplitude snap-through dynamic occurs only in the frequency band from 20 to 25 Hz and the displacement amplitude of intrawell oscillations at higher frequencies is larger. One explanation for this discrepancy is a minor imperfection between the two stable equilibria. Asymmetry of stable equilibria is known to result in displacement amplitudes among the dynamic steady-states that are nearer in value than in the case of perfect symmetry [126] [127]. In addition, the piezoelectric cantilever contains layers of FR4 glass reinforced epoxy laminate substrate, piezoelectric PZT-5H, and copper electrodes. The FR4 layers may possess viscoelastic, and thus time-dependent, material properties that cause more intricate damping influences than the viscous damping model currently employed in the model formulation. These details of experimental system design and implementation may be introduced in a refined model formulation in future investigations. Yet, despite such deviation associated with inevitable limitations to achieve perfect configurations in experiments, the experimental results are in good qualitative and quantitative agreement with the analytical and simulation results, giving validation support to the model formulation and solution efforts.

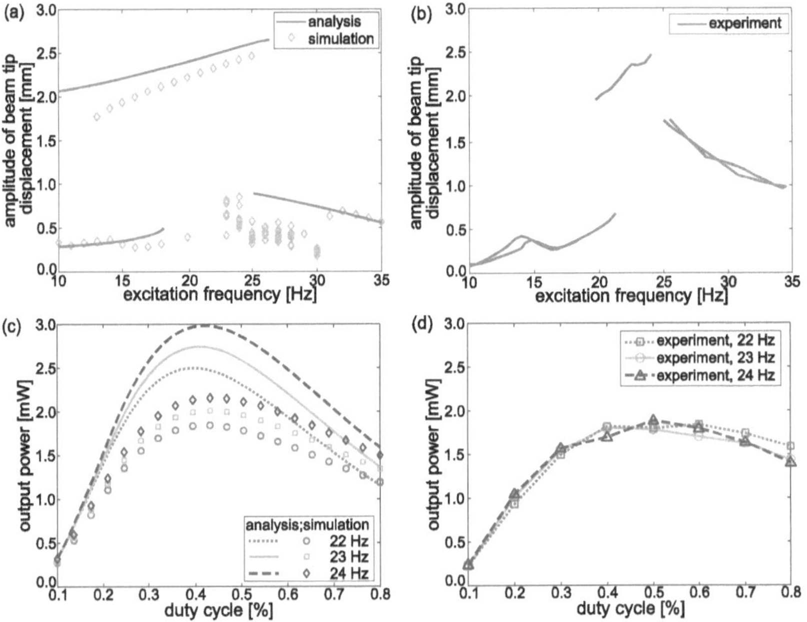



Figure 4.6 Mechanical and electrical response of (a) beam-tip displacement amplitude predicted by analysis and simulation and (c) output power at different excitation frequency given by analysis and simulation; (c) and (d) corresponding results from experiments.

108

Snap-through behavior is predicted by the modeling and experiment in the frequency range from 21 to 24 Hz. To validate the model predictions of the DC power delivery from the buck-boost converter, the frequencies 22, 23, 24 Hz are considered. The output power shown in Figure 4.6(c) and (d) corresponds to the power delivered to the load $R$. The analysis in Figure 4.6(c) predicts the greatest DC power delivery among all results. Because the analysis idealizes the response in steady-state as being from a single harmonic, no higher-order losses are incurred in the analytical prediction that otherwise occur for the simulation and experiment. In addition, both the simulation and the experimental system are influenced by a much more detailed losses of the electrical components, like the on resistances of the diodes and MOSFET that contribute to smaller output powers in simulation and experiment shown in Figure 4.6(c) and (d). Despite this discrepancy, all three methods of assessment identify the same strategy for optimization using the nonlinear energy harvester with buck-boost converter. Specifically, the optimal duty cycles at the three frequencies are all around 0.4% to 0.5%. In addition, a greater reduction of the DC power occurs for duty cycles less than this optimal range than above this range, which is manifest in all of the results.

Overall, the good qualitative and quantitative agreement among all analytical, simulation, and experiments results establishes the efficacy of the analytical model formulation to characterize the mechanical and electrical behavior of the nonlinear energy harvesting system.

**4.5 An impedance-based assessment for optimal DC power delivery**

After validating the analytical method, the same parameters shown in Table 4.2 are taken

to conduct a thorough analysis to reveal the relationships between impedance change and

optimal output power. For the proposed harvesting system, the structural dynamics $x(\tau)$

and electrical voltage across the piezoelectric beam $v_p(\tau)$ are obtained by solving Eq.

(4.22). The corresponding power $\underline{P}_O$ is then determined by Eq. (4.25). Figure 4.7 shows

the output power across the load $R$ at different base acceleration frequency and buck-boost

converter duty cycle. The white dashed line displays the duty cycle at each frequency to

deliver the maximum DC power. A clear frequency dependence of the duty cycle to

maximize the DC power is observed in the analytical results of Figure 4.7.

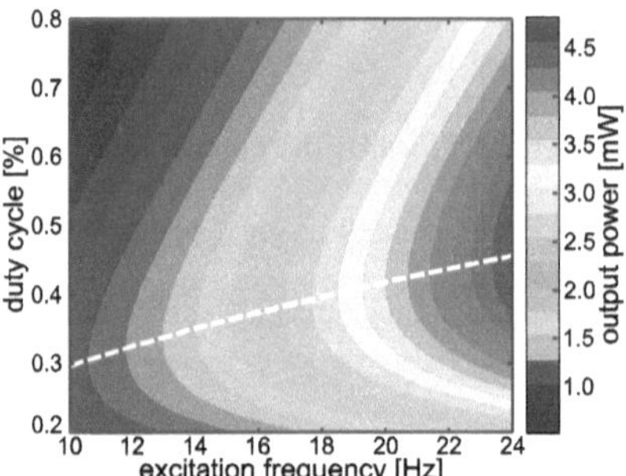

Figure 4.7 Analytical prediction of DC power at base acceleration amplitude 9.8 m/s$^2$

considering the entire nonlinear energy harvester and power management circuit. The

white dashed line shows the duty cycle at each frequency to generate the maximum power.

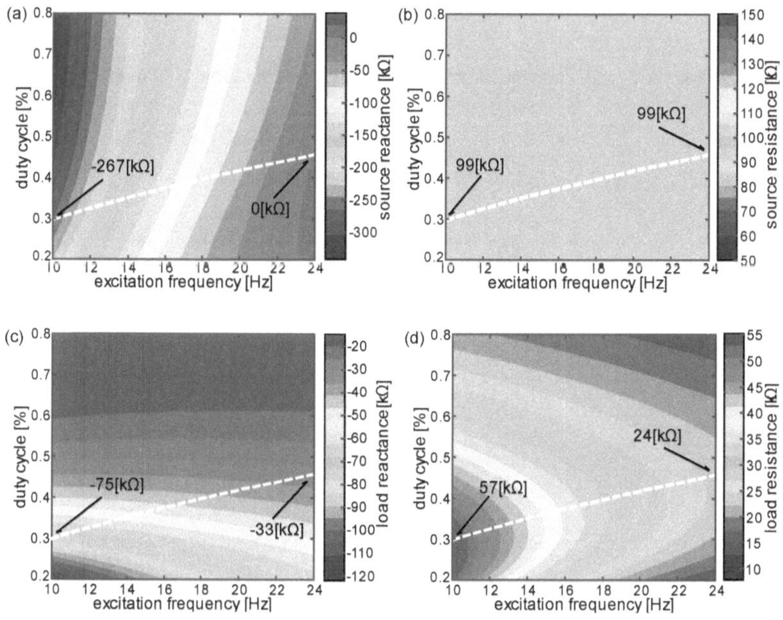

Figure 4.8 Impedance results at base acceleration amplitude 9.8 m/s² of (a) source reactance and (b) source resistance and (c) load reactance and (d) load resistance.

With Eqs. (4.12) and (4.29), the source and load impedances are respectively obtained. Figure 4.8 presents the corresponding impedance results using the same $x$ and $v_p$ values that are computed for the results of Figure 4.7. Figure 4.8(a) and (b) show the influence of change in excitation frequency and duty cycle on the source resistance and reactance, respectively. The source resistance $R_s$ in Figure 4.8(b) is constant at 99 kΩ, as described in Section 4.4. The absolute value of the source reactance $X_s$ decreases with increase in excitation frequency and increases with increase in the duty cycle, Figure 4.8(a). When the

111

source reactance is zero, snap-through behavior is no longer possible. Figure 4.8(c) and (d) present the load impedance results. Although the buck-boost converter displays a pure resistance property, combined with the rectifier bridge and piezoelectric capacitance, the load reactance $X_L$ is influenced by duty cycle and base acceleration frequency. In addition, from Figure 4.8(d) the load resistance $R_L$ presents a maximum value with changing duty cycle at each frequency.

Based on conservation of energy, the power delivered to the load $R$ in Figure 4.5 is equal to the power across $Z_L$ in Figure 4.4(a). Consequently, this output power is

$$P_o = \frac{1}{2} \frac{V_s^2 R_L}{\left(R_S + R_L\right)^2 + \left(X_s + X_L\right)^2} \tag{4.36}$$

By conventional impedance matching theory, the optimal output power is delivered when the load impedance is the complex conjugate of the source impedance. In other words, when $R_L = R_S$ and $X_L = -X_S$. Yet, in the case of this study, the source reactance varies based on the base acceleration frequency and the buck-boost converter duty cycle, as shown in Figure 4.8(a). In addition, the load resistance and reactance are unable to realize the same reactance and resistance values as a conjugate of the source impedance. This is observed by comparing Figure 4.8(c) to (a) and (d) to (b) wherein it is found that the load reactance in Figure 4.8(c) takes on the same sign as the source reactance in Figure 4.8(a). Also, the maximum load resistance in Figure 4.8(d) is less than the source resistance in Figure 4.8(b). In other words, the conventional impedance matching theory is not able to be employed. Yet, despite this limitation, the results of Figure 4.8 indicate that a sub-optimal set of working conditions in the duty cycle and base acceleration frequency are identified that still yield peak output power. The white dashed lines in Figure 4.8 correspond to the full system analysis from Figure 4.7 that results in the optimal working conditions. For Figure 4.8(d), the white dashed line follows the peak value of resistance for a given duty cycle and excitation frequency. Although this value still falls short of the source resistance 99 k$\Omega$, the load resistance is maximized and thus nearest to the ideal condition.

Because the phase of the load impedance is influenced by an intricate combination of power management circuit characteristics, see Eqs. (4.21), (4.23), and (4.30), the load reactance and resistance are unable to be individually selected for the sake of maximizing output power. Based on Eq. (4.36), the influences of load resistance and reactance on the output power are expressed in Eq. (4.37). Here the sum of source and load reactance is taken as a new variable $K = X_S + X_L$.

$$\frac{\partial P_o}{\partial R_L} = \frac{1}{2}V_s^2 \cdot \frac{\left(X_s + X_L\right)^2 - R_L^2 + R_S^2}{\left[\left(R_S + R_L\right)^2 + \left(X_s + X_L\right)^2\right]^2} = \frac{1}{2}V_s^2 \cdot \frac{K^2 - R_L^2 + R_S^2}{\left[\left(R_S + R_L\right)^2 + K^2\right]^2}$$

$$\frac{\partial P_o}{\partial K} = \frac{1}{2}V_s^2 \cdot \frac{-2\left(X_s + X_L\right)R_L}{\left[\left(R_S + R_L\right)^2 + \left(X_s + X_L\right)^2\right]^2} = \frac{1}{2}V_s^2 \cdot \frac{-2KR_L}{\left[\left(R_S + R_L\right)^2 + K^2\right]^2} \qquad (4.37\text{a-b})$$

By setting Eqs. (4.37a,b) to zero and solving for the $R_L$ and $K$, the impedance matching condition is obtained. Considering the parameters of the energy harvesting platform studied here, the comparison of load and source resistance reveals

$$R_S > 2R_L \qquad (4.38)$$

Therefore, the absolute difference of the two gradients is

$$\left|\frac{\partial P_o}{\partial R_L}\right| - \left|\frac{\partial P_o}{\partial K}\right| > V_s^2 \cdot \frac{K^2 + R_L^2 - 2|K|R_L}{\left[\left(R_S + R_L\right)^2 + K^2\right]^2} > 0 \qquad (4.39)$$

The Eq. (4.39) reveals that the change in output power caused by resistance change is more influential than the change in output power that results from the change in reactance. Specifically, when $R_s > 2R_L$ the resistance becomes a more influential factor towards the optimal conditions of duty cycle and excitation frequency more maximum output power. This result explains the fact that the load resistance maxima in Figure 4.8(d) correlate to the peak DC power operating conditions computed from the full analysis (see the white dashed curve in Figure 4.8(d) that is obtained from the maxima in Figure 4.7).

## 4.6 Optimal duty cycle predicted by impedance analysis

From the analysis of the previous section, although the conventional approach to impedance matching is not possible using the nonlinear energy harvesting system considered here, a sub-optimal matching condition still maximizes the output power. To investigate this sub-optimal approach to optimization in further detail, the insight that resistance matching is more influential in the output power is studied further. The Eq. (4.36) is taken omitting the influence of reactance.

$$\underline{P} = \frac{1}{2} \frac{\underline{V}_s^2 R_L}{\left(R_L + R_s\right)^2} \tag{4.40}$$

The equivalent voltage $\underline{V}_s$ and source resistance $R_s$ are constants. In other words, the only variable in Eq. (4.40) is $R_L$, which is related to circuit parameters according to Eq. (4.30).

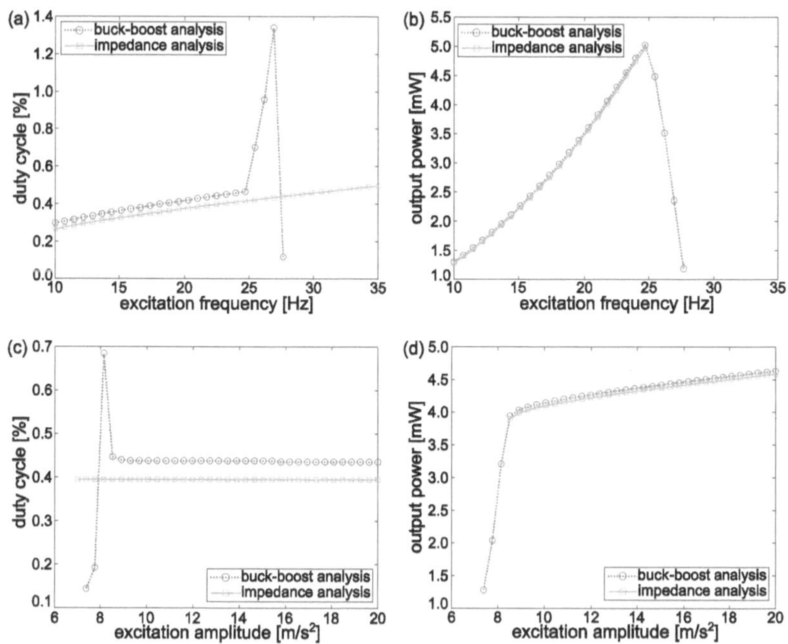

Figure 4.9 (a) Optimal duty cycle of the nonlinear energy harvesting system at various excitation frequencies with a base acceleration amplitude 9.8 m/s$^2$; (b) the optimal power corresponding to (a). (c) Optimal duty cycle at different base acceleration amplitudes with an excitation frequency 22 Hz; (d) the optimal output power corresponding to (c).

The red dotted curves with circle data points in Figure 4.9 represent the results from the analysis considering both the harvester structure and buck-boost converter, termed the 'buck-boost analysis'. The results from the buck-boost analysis are derived from the procedures described in Section 4.3.1. To obtain the results shown in Figure 4.9, using the buck-boost analysis the Eq. (4.22) is solved to identify base acceleration excitation frequency, excitation amplitude, and the duty cycle that lead to optimal working conditions. The green solid curves with square data points in Figure 4.9 correspond to the 'impedance analysis' that is computed using Eqs. (4.21), (4.23), (4.30), and (4.40). From 10 Hz, for increase in the excitation frequency, the optimal duty cycle slightly increases from around 0.3% to 0.5% according to the impedance analysis. Yet, for increase in the excitation frequency around 25 Hz, an abrupt increase in optimal duty cycle is observed in the buck-boost analysis. This large qualitative change in the optimal working conditions of the converter is due to the loss of snap-through dynamics, shown in Figure 4.10. Figure 4.10 presents the (a) amplitude of displacement and (b) resulting output power for the base acceleration frequencies 22, 26, and 27.3 Hz when the base acceleration amplitude is 9.8 $m/s^2$. As the frequency increases, the responses separate into two branches. As observed in Figure 4.10(b), the peak power condition therefore shifts to conditions of higher duty cycle. This trend disagrees with the impedance matching-based results in Figure 4.9(a) that presumes the snap-through behavior may persist for all frequencies. Yet despite this nuance of prediction capability, the output DC powers predicted by the buck-boost and impedance analyses are in good quantitative agreement verifying the viability of the impedance-matching based analysis to characterize the primary trends.

A similar analysis is conducted by studying the role of change in the base acceleration amplitude in Figure 4.9(c) and (d). For amplitudes of the base acceleration from 7 to 9 $m/s^2$, the snap-through dynamic is not possible for all duty cycles for the 22 Hz frequency. This explains the sudden variation in the buck-boost analysis optimal duty cycle that is not evident by the impedance analysis. For base acceleration amplitude greater than 9 $m/s^2$, the optimal duty cycle changes slightly in both analytical approaches. On the other hand, since

the base acceleration amplitude does not affect the load resistance, the change of load

resistance is the same at each of base acceleration amplitude, which explains the constant

optimal duty cycle determined by the impedance analysis. These results establish

confidence in the source impedance analysis.

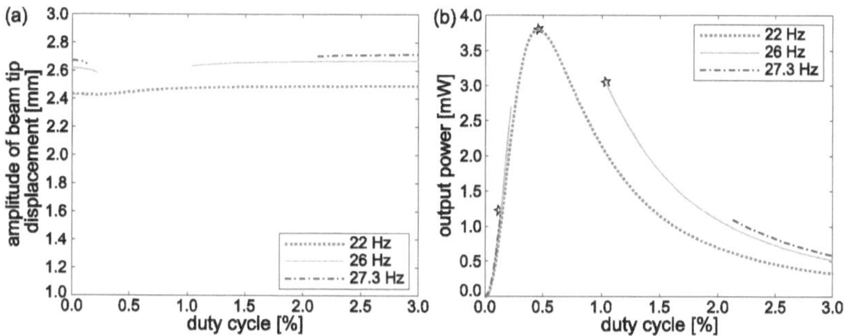

Figure 4.10 (a) Amplitude of beam tip displacement at base acceleration amplitude 9.8

$m/s^2$. (b) Output power as a function of duty cycle at base acceleration amplitude 9.8 $m/s^2$.

Based on the results shown in Figure 4.9, the optimal working condition for the nonlinear

vibration energy harvester interfaced with a buck-boost converter may be determined by

the impedance analysis. In contrast to the buck-boost analysis itself, the impedance analysis

only requires information regarding the source and load resistances. Since both resistances

are derived in Secs. 2 and 3, this research identifies an efficient means by which to evaluate

the optimal working condition of the energy harvesting system.

**4.7 Conclusion**

This research exploits the perspective of impedance to examine the optimal DC power operation of a nonlinear energy harvester interfaced with a buck-boost converter for practical electrical power management. The internal, or source, impedance of the nonlinear harvester platform is shown to be influenced by change in the driven conditions of frequency and amplitude. Because such changes challenge conventional impedance matching concepts, a rigorous analytical study is undertaken to uncover means to maximize DC power delivery from the system despite inability to obtain conventional complex conjugate impedance matching. After validating the analysis, the output power results and impedance results are shown to demonstrate a sub-optimal condition for maximum DC power delivery through the buck-boost converter. By studying the sub-optimal conditions in detail, it is explicitly revealed that resistance of the load plays a more influential role for the system optimization. With this insight, a simplified method of DC power maximization is created that considers only the load and source resistances to optimize the working conditions. Both the predicted duty cycle and output power from such reduced modeling strategy show good agreement with the whole system analysis, exemplifying the strength of the analytical formulation established here upon impedance principles.

## Chapter 5 Investigations on the influences of the energy storage component on output power with SEH, SCE, and parallel-SSHI circuits

Vibration energy harvesting has been seen as a promising energy provider for the emerging Internet of things. Due to the unpredictability of vibration energy, energy storage components are indispensable in an energy harvesting system. Yet, the influences of energy storage components, such as supercapacitor and rechargeable batteries, on the harvested power are seldomly examined in studies. Therefore, in this chapter, a Li-ion rechargeable battery is incorporated with three interface circuits (SEH, SCE, and parallel-SSHI) to scrutinize the principles in maximizing the power delivered from the nonlinear energy harvester. To this end, by employing the analogy between mechanical platforms with electrical components, the nonlinear energy harvesting system can be simplified as a source voltage connecting with a source impedance in series with a load impedance. After the establishment of the source and load impedance models, the essential influence of the battery voltage on maximizing output power is then uncovered. In addition, it is found that when the rechargeable battery is included in the interface circuits, a simplified power calculation solely focused on the load impedance is accurate to estimate the power delivery for various types of interface circuits. After the proposed theoretical approach is verified by the numerical efforts, a comparison among the three interface circuits demonstrates the highest and lowest peak power delivered by interfacing respectively with the parallel-SSHI circuit and SEH circuit. The SCE circuit corresponds to constant output power for all batteries connected. The research sheds light on the influences of energy storage components on the performances of the nonlinear energy harvesting system. The discoveries may help identify the best integration of sub-systems for maximum harvested power when the energy storage component is incorporated.

## 5.1 Introduction

The Internet of Things, or IoT, is a system of interconnected devices to continually collect, process, and share data through a network [71]. The emerging of IoT systems has created strategies to solve long-standing challenges, such as smart transportation, smart healthcare, and smart water [71]. With the growth of the IoT population to 35 billion [128], energy consumption has brought significant challenges in further advancing the technologies [129]. Vibration energy may be a potential energy solution for its abundance and accessibility in the environments that IoT systems are likely deployed [129].

Piezoelectric materials are popularly utilized to convert vibration energy to electrical power [40]. To overcome the narrow-band shortcoming of linear energy harvesters, a piezomagnetoelastic structure is proposed to broaden the effective bandwidth and enhance the harvested electrical power [9] [41] [15] [42]. Because of the coexistence of high and low energy dynamic regimes, various approaches have been established to more readily attain high-energy snap-through vibration [11] [73] [16]. For example, Wang and Liao [44] utilized the load perturbation method to examine strategies that transform the dynamic regime from intrawell to snap-through. Wang and Mann [130] resorted to a reinforcement learning algorithm to study the control policies for co-existing attractor selection.

Despite great advancements achieved, most previous studies performed experiments or applications under predesigned simplified conditions and rarely considered the process of electricity to load terminal [131]. In general, the power required for sensors in an active mode is around a few milliwatts [132]. For a piezomagnetoelastic system running in a high-energy orbit, the harvested electrical power is as high as milliwatt, whereas the low-energy vibration can only provide the power of microwatt [50] [133] [68]. Since the ambient kinetic energy is unpredictable [29], the harvested power from the nonlinear energy harvesting system may be insufficient with a low excitation amplitude even with careful consideration of the mean operating environment. Therefore, an energy reservoir is unavoidable in an IoT system to sustainably provide the required power for the devices and ensure the energy neutral operation that balances the energy harvested and consumed [30]. Supercapacitor and rechargeable battery are two widely explored energy storage options in the vibration energy harvesting community [71]. Yet, according to the survey of state-of-art studies, capacitors are frequently taken as smoothing capacitors to reduce the ripple voltage across the resistance instead of energy storage components. To characterize the performance in charging capacitors, Zhang et al. [134] studied the influence of storage capacitance on maximizing harvested energy with a linear energy harvester connecting to four interface circuits, namely SEH (standard energy harvesting) circuit, SCE (synchronized charge extraction) circuit, parallel-SSHI (synchronized switched harvesting on inductor) circuit, and series-SSHI circuit. Yet, for the investigations on rechargeable batteries, the discussions are mainly focused on piezoelectric beam and harvesting circuits rather than characterizations of the battery itself. Sodano et al. [135] discussed the performance of two types of piezoelectric materials: monolithic piezoelectric and macro fiber composites, in charging nickel-metal hydride batteries. Hu et al. [136] demonstrated enhancements in charging an electrochemical battery by integrating a synchronized switch harvesting on an inductor circuit with a DC-DC converter.

In addition, since the voltage delivered from the piezoelectric beam is an AC signal, a rectification circuit is required to regulate the oscillatory current to DC power for charging a battery. On the basis of a full-wave rectification bridge, advanced harvesting circuits have been created for maximizing electrical energy conversion [113] [114] [115]. For example, parallel-SSHI circuits and SCE circuits respectively add a controlled switch before and after the rectifier bridge to enhance the harvested power [137] [51]. These interface circuits have mostly been analyzed on linear harvesters [134] [113]. Recently, the integration of nonlinearity in both mechanical and electrical platforms has been constructed to examine the overall performance [50] [47] [14] [74]. Yet, characterizations on the performance of charging batteries with nonlinear energy harvesting systems have rarely been seen in the literature.

Considering the state-of-the-art studies, there is a necessity to conduct an analysis to characterize the performances of the nonlinear energy harvesting system in charging a battery. To this end, this report is organized as follows. The next section studies the internal impedance of the nonlinear harvester structural platform. Then, three nonlinear interface circuits, namely SEH, SCE, and parallel-SSHI circuits, are taken into detailed consideration to build the impedance models of the circuits when charging a battery. With the established impedances for the mechanical and electrical platform, a rigorous

122

impedance analysis is proposed to demonstrate the optimal strategies and comparison among the three interface circuits when connecting to a storage component. Finally, a summary of key insights from this work concludes the report.

## 5.2 Impedance model of the nonlinear energy harvester

The detailed impedance model of the nonlinear energy harvester has been presented in Chapter 4. In this section, a simple introduction to the impedance model is included to simplify the discussions in the following discussions.

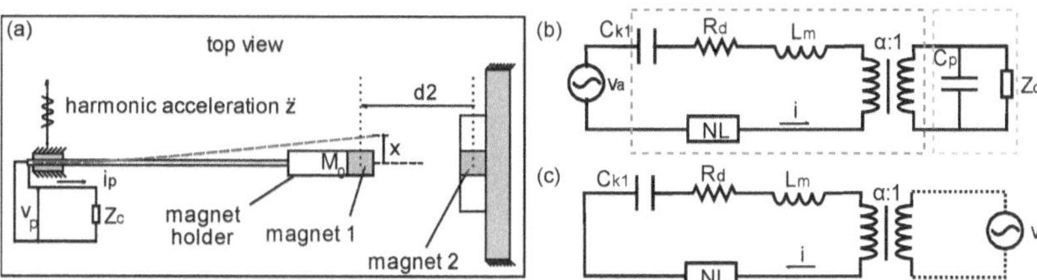


Figure 5.1 (a) Schematic of nonlinear energy harvesting systems. (b) Equivalent circuit of a base-excited nonlinear energy harvesting system. (c) Equivalent circuit for the internal source impedance determination.

The nonlinear vibration energy harvesting system considered in this chapter is shown in Figure 5.1(a). A piezoelectric cantilever has a tip mass $M_0$ constructed by magnet 1 and its holder at the free end. Magnet 2 is installed aside by magnet 1 to introduce nonlinearity into the system by the repulsive magnetic force. With careful adjustment of the distances $d_2$ shown in Figure 5.1(a), the bistable nonlinearity is realized. Since the improvements presented by a bistable vibration energy harvester are associated with the high-energy snap-through behavior, in the following investigations the study places attention on the snap-through response. The beam is excited by a harmonic base motion representative of the ambient kinetic energy. The harvesting circuit is represented by an impedance $Z_c$.

In this work, the frequencies of base acceleration are around the lowest order linear natural frequency of the piezoelectric cantilever, so that the deflection of the cantilever is primarily in the first vibration mode. Therefore, a single degree-of-freedom model is adopted to study the nonlinear energy harvesting system. The governing equations are expressed as

$$m\ddot{x} + d\dot{x} + k_1\left(1-p\right)x + k_3 x^3 + \alpha v_p = -m\ddot{z} \tag{5.1a}$$

$$C_p \dot{v}_p + i_p = \alpha \dot{x} \tag{5.1b}$$

In Eq. (5.1), $x$ is the beam tip displacement relative to the motion of the base displacement $z$ ; $m$, $d$, $k_1$, and $k_3$ are the equivalent lumped mass, viscous damping, linear stiffness, and nonlinear stiffness corresponding to the first vibration mode; $p$ is the load parameter, which is used to indicate the influence of magnetic forces on reducing the linear stiffness; $\alpha$ is the electromechanical coupling constant; $C_p$ is the internal capacitance of the piezoelectric beam; $v_p$ is the voltage across the piezoelectric beam electrodes; $i_p$ is the corresponding current that passes into the harvesting circuit impedance $Z_c$ ; The overdot operator indicates differentiation with respect to time $t$ .

Studies have shown the analogy between mechanical and electrical systems in the energy harvesting literature. Here, such a concept is employed to identify an equivalent electrical system shown in Figure 5.1(b) that represents the vibration energy harvesting system. The electromechanical coupling constant $\alpha$ is interpreted as a transformer turn ratio. In Figure 5.1(b), $v_a$ corresponds to the base acceleration via $v_a = -m\ddot{z}$. The inductance $L_m$, resistance $R_d$, and capacitance $C_{k1}$ in the circuit respectively relate to the mass $L_m = m$, viscous damping $R_d = d$, and compliance $C_{k1} = 1/k_1$. The current through these components is related to the relative velocity of the beam tip via $i = \alpha\dot{x}$. The nonlinearity characterized by the negative linear stiffness $-pk_1$ and nonlinear stiffness $k_3$ can be modeled by a capacitor that has a response-dependent capacitance value, which is represented by a general impedance $NL$ block in Figure 5.1(b) [50].

In the following study, the components in the red dashed box in Figure 5.1(b) are collectively considered to be the source impedance $Z_s$ of the nonlinear energy harvesting system. Correspondingly, the internal capacitor $C_p$ in parallel with the interface circuit impedance $Z_c$ is termed as load impedance $Z_L$. To construct the model of source impedance, the original source $v_a$ in Figure 5.1(b) is removed according to Thévenin's theorem [138]. Instead, an AC voltage $v$ is connected to the output terminal, which is referred to as the driven voltage shown in Figure 5.2. Then, the governing equation for the system shown in Figure 5.1(c) is

$$m\ddot{x} + d\dot{x} + k_1\left(1 - p\right)x + k_3 x^3 = \alpha v \tag{5.2}$$

Here, the driven voltage $v$ is defined as

$$v = -V\cos\left(\omega t\right) \tag{5.3}$$

the solution to (5.2) may be approximated as

$$x(t) = c + a\sin(\omega t) + b\cos(\omega t) \tag{5.4}$$

Using principles of harmonic balance [122], the corresponding mechanical responses can be solved. More details about the responses can be found in the Chapter 4.

Since the current through this source impedance is related to the relative velocity of the beam tip via $i = \alpha \dot{x}$, therefore, the source impedance $Z_s$ can be characterized by Eq. (5.5).

$$Z_s = \frac{v}{i} = \frac{-Ve^{j\omega t}}{\alpha \dot{x}} = \frac{-Ve^{j\omega t}}{\alpha \omega r \cdot e^{j(\omega t - \varphi_i)}} = \frac{-Ve^{\varphi_i}}{\alpha \omega r} \tag{5.5}$$

where

$$\tan \varphi_i = -b/a; \quad r = \sqrt{a^2 + b^2} \tag{5.6 a,b}$$

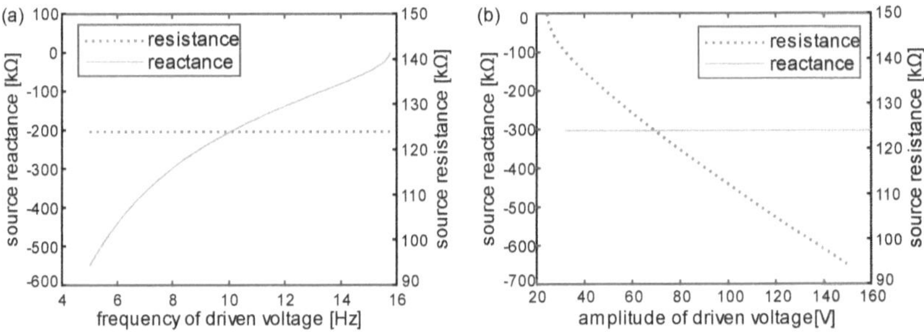


Figure 5.2 (a) Driven voltage frequency and (b) amplitude influences on the source impedance.

Figure 5.2 presents source impedance results that characterize the impedance values associated with the change in the driven voltage. The system parameters used to generate Figure 5.2 are presented in Table 5.1. The amplitude of the driven voltage for Figure 5.2(a) is 50 V and the driven voltage frequency for Figure 5.2(b) is 10 Hz. It is observed from

Figure 5.2 that the reactance is frequency- and amplitude-dependent. Such dependencies are evidence of the response-dependent capacitor representing the cubic nonlinearity and presence of negative linear in the system. In contrast, the resistance is independent of the frequency and amplitude of the driven voltage in Figure 5.1(b).

Table 5.1 Identified system parameters for three nonlinear systems studied in this research.

| $m$ (g) | $p$ (dim) | $k_1$ (N/m) | $k_3$ (MN/m$^3$) | $d$ (N·s/m) | $\alpha$ (mN/V) | $C_p$ (nF) |
|---|---|---|---|---|---|---|
| 18.17 | 1.18 | 556 | 24 | 0.15 | 1.1 | 88 |

After the characterization of the source impedance of the nonlinear energy harvester, in the following sections, the load impedance models for three interface circuits have been established. By leveraging the models of the source and load impedance, the optimal strategies and comparisons among the three circuits have been scrutinized to shed light on maximizing the charging efficiency for the battery.

## 5.3 Impedance analysis for characterizing and comparing performances of interface circuits in charging batteries

Studies have shown improvements in delivering power to resistance by adding a controlled switch to the rectification circuit [52] [137]. Yet, energy storage components are seldomly considered in the investigations, especially when integrated with nonlinear energy harvesters. In this section, three interface circuits are employed to conduct a thorough impedance analysis to demonstrate the approaches in maximizing the energy conversion efficiency for charging batteries with the nonlinear energy harvester.

### 5.3.1 Impedance models for three interface circuits when charging batteries

The three interface circuits considered here are the SEH, SCE, and parallel-SSHI circuits, as shown in Figure 5.3. The electrical components inside the green dashed line determines the load impedance $Z_L$ defined in Section 5.2. The corresponding waveforms of the piezoelectric voltage $v_p$ and the current flows into the battery are presented in Figure 5.4.

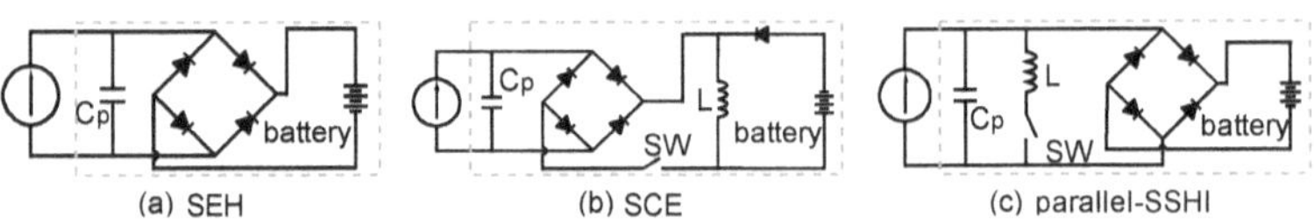



Figure 5.3 Schematics of three interface harvesting circuits for charging battery. (a) SEH circuit. (b) SCE circuit. (c) parallel-SSHI circuit.

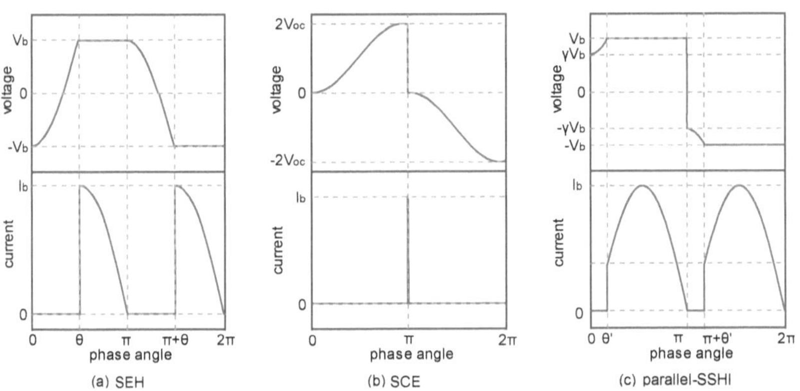



Figure 5.4 Waveforms of piezoelectric voltage and current flow into battery. (a) SEH circuit. (b) SCE circuit. (c) parallel-SSHI circuit.

When charging a battery with an SEH circuit in Figure 5.3(a), a full-wave rectification circuit is directly connected to a rechargeable battery. When the piezoelectric voltage $v_p$ is less than the battery voltage $V_b$, the rectification bridge is reverse biased, the current through the battery is zero as in Figure 5.4(a). Otherwise, the rectification bridge is conducted to charge the battery with a current $i_b$, and the piezoelectric voltage is sustained at the battery voltage $V_b$. The phase angle $\theta$ in Figure 5.4(a) defines the block angle of the diode bridge, which is determined by Eq. (5.7). Since the rectified voltage of the SEH circuit is confined by the battery voltage $V_b$, by approximating the piezoelectric voltage with its fundamental harmonic term, the load impedance and the amplitude of the dominant harmonic component of the piezoelectric voltage are expressed as a function of battery voltage and open-circuit voltage according to Eqs. (5.7) , (5.8), and (5.9) for the interface SEH circuit [26] .

$$\cos\theta = 1 - 2V_b/V_{oc} \tag{5.7}$$

$$Z_{SEH} = \frac{1}{\pi\omega C_p}\left[\sin^2\theta + j\left(\sin\theta\cdot\cos\theta - \theta\right)\right] \tag{5.8}$$

$$V_{p,f}^{SEH} = \frac{V_{oc}}{\pi}\sqrt{\sin^4\theta + (\sin\theta\cdot\cos\theta - \theta)^2} \tag{5.9}$$

where $V_{oc}$ is the open-circuit voltage amplitude that can be determined by solving Eq. (5.1) with $i_p = 0$ [122]; the real and imaginary parts of $Z_{SEH}$ respectively correspond to the load resistance and reactance for the system in Figure 5.3(a); $V_{p,f}^{SEH}$ is the amplitude of the fundamental harmonic of the piezoelectric voltage.

Comparing with the SEH circuit, a switch and inductor are added in series after the rectification bridge with a diode and battery connecting in parallel with the inductor to construct the SCE circuit in Figure 5.3(b). The switch $SW$ is controlled to conduct only when the displacement $x$ reaches its peaks. As such, the operation of the energy harvester interfaced with the SCE circuit can be seen as three separate processes. First, the voltage accumulates across the capacitor $C_p$ to reach its peaks. Then the switch $SW$ is closed for a time determined by the $C_p$ - $L$ circuit to transfer the energy stored in the capacitor $C_p$ to the inductor $L$. Meanwhile, the piezoelectric voltage $V_p$ drops from $2 V_{oc}$ to zero as in Figure 5.4(b) [51] [139]. Finally, the electrical energy saved in the inductor $L$ flows into the battery instantaneously as demonstrated in Figure 5.4(b). Given the nature of the SCE circuit, the charge removed from the piezoelectric beam is the same for a determined excitation condition despite the battery connected. Therefore, the load impedance expression for the SCE circuit in parallel with the capacitor $C_p$ inside the green rectangular in Figure 5.3(b) is a unique value for an angular frequency $\omega$ as shown in Eq. (5.10) [139].

$$Z_{SCE} = \frac{1}{\omega C_p}\left(\frac{4}{\pi} - j\right) \tag{5.10}$$

The corresponding amplitude of the fundamental harmonic of the piezoelectric voltage is

$$V_{p,f}^{SCE} = V_{oc}\sqrt{1+16/\pi^2} \tag{5.11}$$

The parallel-SSHI circuit introduces a controlled inductor in parallel with the capacitor $C_p$ in front of the rectification bridge. Similar to the SCE circuit, the switch is controlled to conduct only when the displacement response achieves its extreme values. After the conduction of the switch, the piezoelectric voltage reverses from $V_b$ to $-\gamma V_b$. Here, the inversion factor $\gamma$ relates to the quality factor of the switching loop $C_p - L$ and satisfies (-1,0] limits [140]. Therefore, for a parallel-SSHI circuit, the rectification bridge is open-circuited at first. When the piezoelectric voltage is accumulated to $V_b$, the rectification bridge is then conducted to charge the battery with the current $i_b$ in Figure 5.4(c). The block angle of the diode bridge $\theta'$ is defined in Eq. (5.12). Similar to the SEH circuit, the block angle can be expressed as a function of the battery voltage and open-circuit voltage. The load impedance when interfacing with the parallel-SSHI circuit is given in Eq. (5.13) [26].

$$\cos\theta' = 1 - (1+\gamma)V_b/V_{oc} \tag{5.12}$$

$$Z_{p\_SSHI} = \frac{1}{\pi\omega C_p}\left[(1-\cos\theta)\left(\frac{4}{1+\gamma}-1+\cos\theta\right)+j(\sin\theta\cos\theta-\theta)\right] \tag{5.13}$$

The amplitude of the fundamental harmonic piezoelectrical voltage is presented below.

$$V_{p,f}^{SSHI} = \frac{V_{oc}}{\pi}\sqrt{(\frac{4}{1+\gamma}-1+\cos\theta)^2+(\sin\theta\cdot\cos\theta-\theta)^2} \tag{5.14}$$

Based on the above discussions, when incorporating a rechargeable battery in the interface circuits, the load impedance and piezoelectrical voltage are determined by the open-circuit voltage and battery voltage for a given interfacing circuit. As such, the investigations on the electrical responses can be examined by studying the mechanical and electrical sub-system separately. In the following section, the attention is solely on the evaluation of the electrical subsystem to scrutinize the influences of the rechargeable battery on power delivered with the provided open-circuit voltages. The findings may help determine the optimal strategies for charging batteries with the nonlinear energy harvester interfacing with various types of circuits in a more efficient way.

### 5.3.2 Comparison among the load impedance for three interface circuits

Because kinetic energy is unpredictable [29], in order to continuously provide the required energy for electronic devices an energy storage component is necessary. In Section 5.3.1, the load impedance models for three interface circuits when charging a battery are constructed. In this section, examples of the load impedances identified for the three interface circuits are presented to illustrate the characterizations of each circuit in terms of the load impedance.

Based on the discussion in Section 5.3.1, the load impedances are associated with the open-circuit voltage and the battery voltage. The excitation condition of case 2 in Table 5.5 is taken as an example to plot the load impedances for three circuits in Figure 5.5. For the parallel-SSHI circuit, the influence of the inversion factor $\gamma$ on the load impedance is also considered in Figure 5.5.

Table 5.2 Excitation frequency and excitation amplitude.

| Excitation condition | Case 1 | Case 2 | Case 3 | Case 4 |
|---|---|---|---|---|
| Excitation frequency [Hz] | 8 | 10 | 12 | 14 |
| Excitation amplitude [m/s$^2$] | 7 | 7 | 10 | 10 |
| Open circuit voltage [V] | 40.70 | 41.57 | 45.65 | 48.25 |

As demonstrated in Figure 5.5, the load impedance curve corresponding to the SEH circuit is cut-off around 41 V of the battery voltage. This is because the open-circuit voltage for the excitation condition in case 2 is 41. 57 V as in Table 5.2. When the battery voltage is higher than 41.57 V, the rectification bridge in Figure 5.3(a) is reverse biased to stop charging the battery. Therefore, the SEH circuit cannot provide energy for batteries with a voltage higher than the open-circuit voltage of the energy harvester. In addition, in Figure 5.5(a), the load reactance associated with the SEH circuit reaches its maximum when the battery voltage equals the open-circuit voltage, whereas the maximum resistance is at the battery voltage that is half of the open-circuit voltage based on Eqs. (5.7) and (5.8).

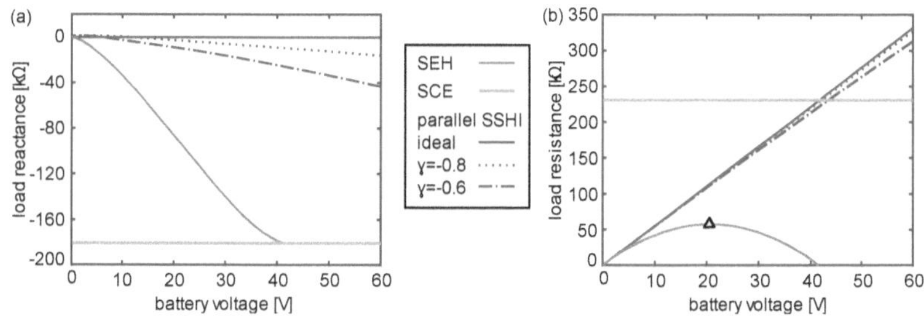


Figure 5.5 Load impedance plots when interfacing with three circuits as a function of battery voltage. (a) Load reactance. (b) Load resistance.

For the SCE circuit, a constant load impedance is presented in Figure 5.5 for all battery voltages according to Eq. (5.10), and the load reactance is the maximum value of the SEH circuit as in Figure 5.5(a). According to the operation of the SCE circuit described in Section 3.1, the battery is charged with the electrical energy stored in the inductor after the rectification bridge. The energy transition between the capacitor $C_p$ and the inductor $L$ is not blocked by the rectification bridge, such that the SCE circuit is possible to charge a battery with any nominal voltage.

133

Moreover, for the parallel-SSHI circuit, three inversion factors are considered to investigate the influence of the inversion factor on the energy conversion efficiency. In Figure 5.5(b), the ideal parallel SSHI circuit is associated with $\gamma = -0.98$. The other two inversion factors are respectively $\gamma = -0.8$ and $\gamma = -0.6$. As shown in Figure 5.5, the absolute values of load reactance and resistance related to the parallel-SSHI circuit respectively increase and decrease with the increase of the inversion factor $\gamma$. Based on the load impedance given in Eq. (5.13), the maximum load reactance with the parallel-SSHI circuit equals the load reactance of the SCE circuit, yet associates with a much higher battery voltage according to the definition of the rectification bridge block angle in Eq. (5.12). For example, for the inversion factor $\gamma = -0.8$, the load reactance reaches its maximum value at battery voltage 181 V, which is also the cut-off voltages for the parallel-SSHI circuit when charging a battery. Figure 5.5 only shows the load impedance corresponding to the battery voltage from 0 to 60 V to demonstrate a much faster increase in load resistance and a slower increase in the absolute value of the load reactance with the increase of the battery voltage.

Overall, comparing the three interface circuits, the highest and lowest load reactance are respectively associated with the SCE and parallel-SSHI interface circuits. For the load resistance, when interfacing with the SEH circuit the system possesses the smallest resistance value, whereas the load resistance for the parallel-SSHI circuit keeps increasing to a much higher value with the battery voltage until the cut-off voltage is reached. In addition, the investigations on the load impedance indicate the existence of cut-off voltage defined by the block angle of the rectification bridge when utilizing the SEH circuit or the parallel-SSHI circuit to charge the battery.

### 5.3.3 Impedance analysis for maximizing charging efficiency

Since both the source and load impedance have been constructed, the nonlinear energy harvesting system in Figure 5.1(b) can be simplified to a voltage source $v_a$ connecting with a source impedance $Z_S$ in series with a load impedance $Z_L$. Therefore, the power harvested by the load $Z_L$ can be determined by Eq. (5.15).

$$P_o = \frac{1}{2} \frac{V_a^2 R_L}{\left(R_S + R_L\right)^2 + \left(X_s + X_L\right)^2} \tag{5.15}$$

In Eq. (5.15), $V_a$ is the amplitude of the source voltage $v_a$ in Figure 5.1(b), $R_s, R_L, X_s, X_L$ are respectively the source resistance, load resistance, source reactance, and load reactance. According to the discussions in Section 5.2, the source resistance is a constant 120 kΩ as in Figure 5.2. The source reactance is regulated by the voltage frequency and amplitude applied across the electrical components in the red box in Figure 5.1(b). As such, the conventional impedance analysis from Eq. (5.15) is intricate to scrutinize the optimal strategies of three interface circuits in charging batteries.

Based on the discussion in Section 5.3.1, the electrical responses when charging batteries can be determined by separately examining the mechanical and electrical platforms. In Section 5.3.1, the dominant harmonic piezoelectric voltage amplitude and the load impedance are provided through Eq. (5.7) to Eq. (5.14) by simply evaluating the interface circuits in parallel with the capacitor $C_p$. The harmonic balance method can then be employed to calculate the open-circuit voltage for the piezomagnetoelastic beam [122].

Therefore, the electrical response applied to the load impedance can then be completely determined. For the simplified circuit, the converted electrical power can also be estimated by Eq. (5.16).

$$P_o = \frac{1}{2} \cdot V_{p,f}^2 \cdot \frac{R_L}{\left(R_L\right)^2 + \left(X_L\right)^2} \tag{5.16}$$

where $V_{p,f}$ represents the amplitude of the fundamental harmonic term of the piezoelectric voltage.

With the excitation conditions given in Table 5.2, the harvested power calculated from Eq. (5.16) for three interface circuits is presented in Figure 5.6, which is marked as analytical results. To validate the estimations, the equations of motion in Eq. (5.1) are simulated via a MATLAB Simulink model by the equivalent circuit in Figure 5.1(b). Runge-Kutta numerical integration is employed for any given base acceleration combination of amplitude and frequency. The calculation time for each harmonic excitation condition is over 500 periods to ensure the steady-state is reached. The simulated power as a function of the battery voltage is taken as a comparison to the analytical estimations for the three interfacing circuit under the excitation conditions in Table 5.2.

Comparing the simulations with the analysis in Figure 5.6, the simplified power estimation by separately evaluating the mechanical and electrical systems can be utilized to capture the harvested power for the SEH circuit and SCE circuit. For the ideal parallel-SSHI circuit, the piezoelectrical voltage possesses a square waveform that can be represented by its fundamental harmonic component. Yet, with the increase of the battery voltage or the decrease of the inversion factor, the piezoelectric voltage may be better represented by a summation of multiple harmonic frequencies. As such, the harvested power characterized

by the fundamental harmonic term deviates from the simulated power for the parallel-SSHI

circuit with a high inversion factor at large battery voltage.

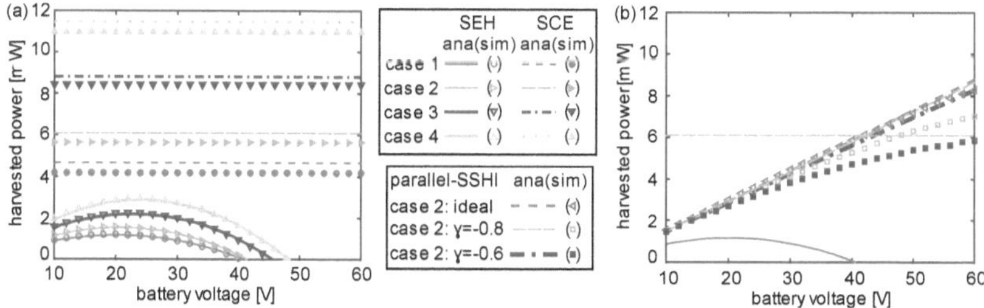

Figure 5.6 Analytical and numerical results for harvested power with (a) SEH and SCE interface circuits and (b) parallel-SSHI circuits as a function of rechargeable battery voltage.

Moreover, according to Figure 5.6(a), the optimal power is always associated with half of

the open-circuit voltage for the SEH circuit, which is also the location of the maximum

resistance for the SHE circuit identified in Figure 5.5(b). This is mainly due to the confined

resistance value associated with the SHE circuit as demonstrated in Section 5.3.2. Eqs.

(5.17) and (5.18) respectively display the derivatives of harvested power regarding the load

resistance and the sum of the source and load reactance. Here, $K$ represents the summation

of the source reactance $X_S$ and load reactance $X_L$.

$$\frac{\partial P_o}{\partial R_L} = \frac{1}{2} V_a^2 \cdot \frac{K^2 - R_L^2 + R_S^2}{\left[\left(R_S + R_L\right)^2 + K^2\right]^2} \tag{5.17}$$

$$\frac{\partial P_o}{\partial \mathrm{K}} = \frac{1}{2} V_a^2 \cdot \frac{-2\mathrm{K}R_L}{\left[\left(R_S + R_L\right)^2 + \mathrm{K}^2\right]^2} \tag{5.18}$$

According to the derivatives in Eqs. (5.17) and (5.18), when the source resistance is much larger than the load resistance, the load resistance provides an essential influence on the output power. The study revealed that $R_L < R_s / 2$ is a sufficient condition to only consider the crucial influences of the resistance in maximizing the power conversion [50]. Given the load resistance shown in Figure 5.5(b), only the load resistance when interfacing with the SEH circuit is much smaller than the source resistance. As such, the optimal power for the SEH circuit is associated with the maximum load resistance for its minimum distance away from the source resistance.

For the SCE circuit, the output power is unchanged for all battery voltages at a given excitation condition as presented in Figure 5.6(a). Moreover, for the given weak-coupling energy harvesting systems, the output power for the SCE circuit is around 4 times of the SEH circuit as manifested in Figure 5.6(a) [141].

For the parallel-SSHI circuit, the harvested power keeps increasing with the increase of the battery voltage to 60 V, which indicates a higher power harvested from the nonlinear energy harvester when charging a battery with higher voltage. Yet, the harvested energy cannot keep increasing with the battery voltage. Figure 5.7 presents the mechanical response for the ideal parallel-SSHI circuit when charging a battery of 70 V under the excitation given in case 2. As shown in Figure 5.7, the higher output power may interfere with the high-energy vibrations to result in an aperiodic oscillation, which suggests more attention should be paid in selecting the interfacing circuits to balance the enhancements achieved by introducing nonlinearity in both sub-systems.

In addition, since the reactance of the SCE circuit is much larger than the reactance of the ideal parallel-SSHI circuit according to Figure 5.5, the maximum power harvested from the ideal parallel-SSHI circuit is larger than the power delivered from the SCE circuit in Figure 5.6(b). On the other hand, given the impedance variation with the inversion factor $\gamma$ presented in Figure 5.5(b), the harvested power for the parallel-SSHI circuit is decreased with the increase of $\gamma$ due to an increase in load reactance. From Figure 5.6(b), the out-performance of the parallel-SSHI circuit at higher battery voltages over the SCE circuit reduces with the increase of the inversion factor.

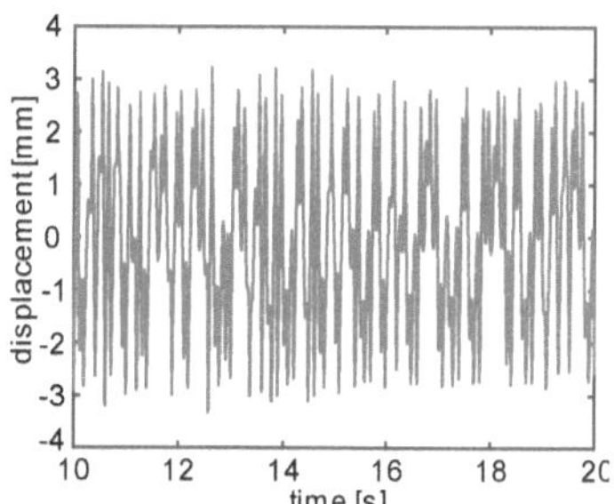

Figure 5.7 Displacement time series for the ideal parallel-SSHI circuit in charging a battery of 70 V under the excitation conditions of case 2.

Through the investigations, when charging a battery with a nonlinear energy harvesting system the electrical responses can be determined by examining the structures and circuits

139

individually. The proposed simplified impedance analysis can provide an excellent estimation on the converted power for three interface circuits to help understand the principles in selecting the interface circuit and maximizing the power delivery.

## 5.4 Conclusion

This research conducts a simplified impedance analysis for the nonlinear energy harvester interfacing with the SEH circuit, SCE circuit, and parallel-SSHI circuit to examine the performances of the three circuits in charging batteries. An analogy between the mechanical platform and electrical components has been introduced to characterize the source impedance. After the establishment of the load impedance models for the three circuits charging batteries, the strategies for maximizing the energy conversion efficiency of the three nonlinear energy harvesting systems have been examined and compared from the perspective of impedance. The findings overall suggest the significant influences of the battery voltage on extracting power from the nonlinear energy harvesting systems. The highest power can be harvested by employing a parallel-SSHI circuit for the smallest reactance. The SCE circuit provides constant output power over all batteries for its unique load impedance. Because of the smallest load resistance among the three circuits, the peak power harvested with the SEH circuit is the lowest with battery voltage at half of the open-circuit voltage. These findings may help the selection of the interface circuit for charging a battery with a nonlinear energy harvester to construct the best integration of sub-systems.

**Chapter 6 Vibration energy harvesters with optimized geometry, design, and nonlinearity for robust direct current power delivery**

With an ever-growing internet-of-things, vibration energy harvesting has attracted broad attention to replace consumable batteries to power the many microelectronic devices. To this end, an energy harvester must deliver the required power to an electrical load over a long time horizon. Yet, design practices for energy harvesters often report strategies based on maximizing output voltage and wide frequency range of operation, which is not directly related to performance-robust functioning. Motivated to provide valuable insight to practical development of vibration energy harvesters, this research develops an analytical modeling framework and optimization technique to guide attention to piezoelectric laminated energy harvesting cantilevers with balanced and robust performance characteristics. The model is numerically and experimentally validated to confirm the efficacy of the optimization outcomes. The results indicate that laminated trapezoidal beam shapes with monostable configuration are the best solution to broaden the frequency range of enhanced dynamic behavior, minimize strain at the clamped beam end, and maximize the output voltage in a rectifier circuit. The results also find that the selection of tip mass may not be highly influential for the overall performance so long as the beam shape, beam length, and placement of nonlinearity-induced magnets are appropriately chosen.

## 6.1 Introduction

With an accelerating development of the internet-of-things (IoT), sustainably powering the low-power wireless devices is on the verge of crisis [141]. With a projected growth of IoT devices upwards of tens of billions by 2020 [142], the excessive reliance on chemical batteries threatens the environment and resilience of the global energy economy. Yet, vibrational energy harvesting suggests a promising solution to meet a portion of the accelerating power supply demand [143]. Optimizing such vibration energy harvesters to ensure robust DC power when subjected to broadband vibration excitation, and over a long time horizon, is therefore central to the vision of a sustainable IoT [144].

The fundamental requirement of a vibration energy harvester is to ensure the necessary DC power is delivered to an electrical load. Optimization strategies therefore seek to enable this requirement. A straightforward method to maximize output power is to increase the electromechanical coupling coefficient for a piezoelectric harvester platform, which enhances the conversion efficiency between vibration and electrical energies. Cho et al. [33] uncovered the influences of residual stress, layering thicknesses, and electrode coverage on the resulting electromechanical coupling coefficient for an optimized thin-film PZT membrane design. Wang and Wu [34] examined the effects of the piezoelectric patch positioning and size for application on a cantilever beam, to show how power harvesting efficiency could be optimized. Lü et al. [35] proposed an optimization method by introducing an intrinsic power density scale to reduce the number of parameters for optimizing energy conversion efficiency. Qin et al. [145] investigate the influence of dimensions on the PZT-5A material and arrived at an optimal electromechanical coupling coefficient ( $k_{15}$ ) for the shear vibration mode of energy harvesting. In addition, after identifying the superiority of a trapezoidal piezoelectric beam shape for mechanical integrity and long service life [146], alternative tapered beam shapes have been studied to enhance the uniformity of strain distribution along the beam length [45] [147] [148]. Dietl and Garcia [32] developed an optimization tool on the basis of such trends and proposed an optimal curved piezoelectric beam shape for more uniform strain distribution. Further approaches to the beam configuration have been considered for sake of maximizing electrical power generation, for example a tapered beam with cavity [149], right-angle piezoelectric cantilevers having auxiliary beams [150], beams with initial curvature [151], and piezoelectric energy harvesters having cellular honeycomb structures [152].

Because linear dynamic response only ensures high output power around the primary resonant frequencies, robust electrical power delivery must involve an energy harvester exhibiting large amplitude dynamic behaviors when subjected to broadband frequency vibration excitation. One well-known method to broaden the frequency range of a piezoelectric energy harvester is to introduce nonlinear magnetic effects [9] [122]. Erturk and Inman [42] suggested as much as an 800% increase in output power could be achieved for a bistable magnetopiezoelastic harvester under harmonic vibration when the platform was interfaced with a purely resistive electrical load. Comparatively, Ferrari et al. [15] reported a 250% increase in output power for a magnetoelastic energy harvester subjected to wideband stochastic excitation. The frequency range for large amplitude response and large output power has been extended also by adopting adaptive bistable harvester designs to more often realize the snap-through behavior [153] [154] [17]. In this spirit, Zhou et al. [16] proposed a flexible bistable energy harvester with two elastic beams to create a variable repulsive magnetic force, which may reduce the potential barrier that must be overcome to ensure snap-through oscillation. Recently instead of focusing on the bistable system, multi-stable systems, such as tri-stable energy harvester, have been introduced to enhance the frequency range and output power achieved for the large amplitude dynamic behaviors [155] [156]. Because the AC output from the harvester must be rectified to DC voltage to power electronic devices, nonlinear harvester platforms interfaced with rectifier circuits [14] or power management circuits [50] have also demonstrated broadband frequency response despite interfacing with such nonlinear electrical circuits.

Yet, enhanced dynamic behavior may correspondingly be detrimental to the long working life of the vibration energy harvester. In particular, the PZT material is exceedingly brittle such that it is vulnerable to damage caused by events resulting in high bending strain. Without cautious design, a piezoelectric energy harvester could fail in hours, or even in minutes [157] [158] [159]. Gundimeda et al. [160] have helped quantify the mechanical integrity and design flexibility of laminated beams when compared with conventional metal substrates used for piezoelectric beams. Li et al. [161] further demonstrated the superiority of laminated piezoelectric beams to yield high output power. Yet, strain conditions or laminated beams are seldom considered when optimizing the energy harvester, especially if nonlinearity is introduced.

Given the state-of-the-art in optimization efforts for piezoelectric energy harvesters, a need exists to develop an optimization approach that exploits nonlinearity in the design of custom-shaped harvesters having excellent DC output power, broad frequency range of operation, and exceptional mechanical integrity. This research meets the need via an analytical model and integrated genetic algorithm optimization framework. The following sections introduce the analytical method, numerically verify the efficacy of the analysis, and then establish the optimization framework. Optimized piezoelectric energy harvesters are then examined and cross-compared, after which an experimental sequence confirms the validity of the theoretical predictions through laboratory examinations. A summary of main discoveries from this research are presented to conclude this report.

## 6.2 Analytical model overview

### 6.2.1 Analytical model formulation for a laminated piezoelectric beam with magnetic nonlinearity

Figure 6.1 presents a schematic of the nonlinear energy harvesting system considered in this research. A laminated piezoelectric cantilever is interfaced with a rectifier bridge $D_i$, smoothing capacitor $C_L$, and load resistance $R$, which is a standard energy extraction circuit. The lamination sequence is intricate. As shown in Figure 6.1(b), along the total length $L$ of the clamped beam, the width at the free end $b_L$ and the width at the fixed end $b_0$ are the layers D, composed from a glass-reinforced epoxy laminate (FR4). The thickness of the FR4 is $2h_s$. Since ceramic materials are exceedingly difficult to arbitrarily shape due to extreme brittleness, here we consider a piezoelectric PZT-5H layer labeled as layer B that adopts a fixed and commercially available rectangular shape. The subsequent optimization of this report considers other beam design parameters as candidates for optimization, in contrast to change of the PZT-5H layer shape. The length, width, and thickness of the piezoelectric layer considered here are $L_p$, $b_p$, and $h_p$. The two piezoelectric layers shown in Figure 6.1(c) are connected in parallel. The layer A corresponds to a copper electrode layer and has the same length and width as the piezoelectric layer for full electrode coverage. The thickness of the copper layer is $h_c$. Layer

145

C denotes polyimide that fills the remaining laminate beam volume in the absence of the copper or piezoelectric layers. Repulsive magnets are included to introduce nonlinearity for tuning the frequency response when the cantilever is subjected to the harmonic base acceleration $\ddot{z}$. The total tip mass is taken as $M_0$, which constitutes both the magnet as well as the holder required to secure the magnet to the cantilever free tip. The other opposing repulsive magnet is attached to the moving base. In the following derivation, the magnet holder shown in Figure 6.1(a) is assumed to be a rigid extension from the beam tip. The length of the holder is $d_1$. The distance between two magnet centers is $d_2$.

When the transverse displacement at the free end of the beam is $w(x,t)$, the strain distribution including the nonlinearity [162] caused by the large amplitude vibration is written

$$S_1 = -z\left[w_{xx}\left(1-w_x^2\right)^{-1/2}\right] \simeq -zw_{xx}\left(1+\frac{1}{2}w_x^2\right) \tag{6.1}$$

The $w_x$ represents for $\dfrac{\partial w}{\partial x}$, $w_{xx}$ is the corresponding second derivative, $z$ is the distance away from the neutral axis, and $S_1$ indicates the strain in the $x$ direction caused by deflection in the $z$ axis.

For the piezoelectric layers, the electrical potential $\varphi(z,t)$ is assumed to be a linear function through the respective piezoelectric layer thickness [163] [164]. Therefore,

$$E_3 = -\frac{\partial \varphi}{\partial z} = -\varphi_z \tag{6.2}$$

where $E_3$ is the electric field in the $z$ direction. Then, the coupling between mechanical and electrical responses for the piezoelectric layers are

$$
\begin{bmatrix} D_3 \\ T_1 \end{bmatrix} = \begin{bmatrix} \varepsilon_{33}^s & e_{31} \\ -e_{31} & E_p \end{bmatrix} \begin{bmatrix} E_3 \\ S_1 \end{bmatrix}
$$

$(6.3)$

The $D_3$ is the electric displacement through the piezoelectric layer thickness, $T_1$ and $S_1$ are the stress and strain in $x$ direction, $e_{31}$ refers to the coupling between the electric field in the $z$ direction and the stress in the $x$ direction, $\varepsilon_{33}^s$ is the piezoelectric permittivity at constant strain, $E_p$ is the Young's modulus of the PZT-5H.

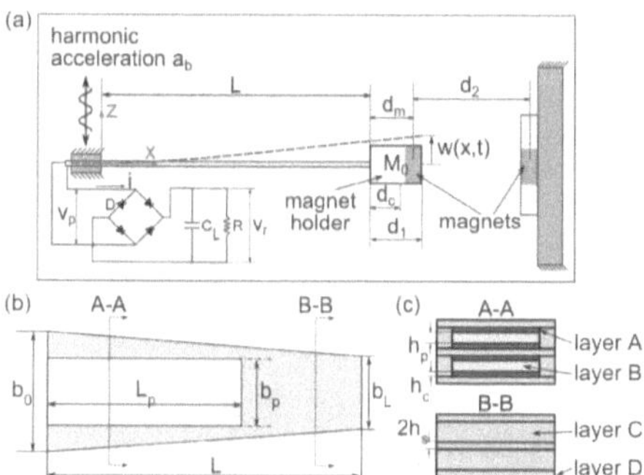


Figure 6.1 (a) Schematic of nonlinear energy harvester system with repulsive magnets and rectification circuit interface. (b) Dimensions of the piezoelectric beam. (c) Cross-section of the piezoelectric beam among the several layers.

Since the rigid magnet holder has a finite length, the kinetic energy of the tip mass $T_m$ is calculated based on the velocity at the mass center. The distance from the laminated beam end to the mass center of the holder is $d_c$ as shown in Figure 6.1(a).

$$
T_m = \frac{1}{2} M_0 \left[ \dot{w}(L+d_c,t) + v_b \right]^2 = \frac{1}{2} M_0 \left[ \dot{w}(L,t) + d_c \cdot \dot{w}_x(L,t) + v_b \right]^2
$$

$(6.4)$

The $\dot{w}$ is the first derivative of the displacement with respect to the time $t$, $v_b$ is the velocity of the base.

Therefore, the total kinetic energy of the nonlinear system shown in Figure 6.1(a) is given in (6.5).

$$T = \frac{1}{2}\iiint_{V_s} \rho_s \left[ \dot{w}(x,t)+v_b \right]^2 dV + \frac{1}{2}\iiint_{V_p} \rho_p \left[ \dot{w}(x,t)+v_b \right]^2 dV + \frac{1}{2}\iiint_{V_c} \rho_c \left[ \dot{w}(x,t)+v_b \right]^2 dV +$$
$$\frac{1}{2}\iiint_{V_e} \rho_e \left[ \dot{w}(x,t)+v_b \right]^2 dV + T_m \tag{6.5}$$

Here and after, the subscripts $s$, $p$, $c$, and $e$ refer to the material of FR4, PZT-5H, copper, and polyimide. These subscripts are then used for the following parameters. The $\rho$ and $V$ denote the corresponding material density and the total volume for each material.

The magnetic potential energy caused by the magnets [9] [15] [165] is

$$U_m = -\int F_{mag}\, dz = -\frac{1}{2}k_1 w^2 \left(L+d_{mc},t\right) - \frac{1}{4}k_3 w^4 \left(L+d_{mc},t\right) \tag{6.6}$$

where $d_{mc}$ is the distance between the magnet center installed on the beam to the laminated beam end.

The magnet pair shown in Figure 6.1(a) include identical magnets. As such, the coefficients in (6.6) are

$$k_1 = f_0 d_2^{-5}, \quad k_3 = -2.5 f_0 d_2^{-7}, \quad f_0 = 3\tau m^2/(2\pi), \quad m = M_a \cdot V_{mag} \tag{6.7}$$

The $\tau$ is the permeability constant in vacuum, $m$ is the effective magnetic moment, $d_2$ is the center-to-center distance between the two repulsive magnets, $M_a$ is the magnetization of the magnet, $V_{mag}$ is the magnet volume.

The total potential energy for the system is supposed to be

$$U = \iiint_{V_s} \frac{1}{2} S_1 T_1 dV + \iiint_{V_c} \frac{1}{2} S_1 T_1 dV + \iiint_{V_p} \frac{1}{2} S_1 T_1 dV + \iiint_{V_e} \frac{1}{2} S_1 T_1 dV - \iiint_{V_p} \frac{1}{2} E_3 D_3 dV + U_m \tag{6.8}$$

where the term $T_1$ refers to the stress in the $x$ direction.

Substituting Eqs. (6.1), (6.3), and (6.6) into Eq. (6.8), the total potential energy is then given by

$$U = \iiint_{V_s} \frac{1}{2} E_s \left[ z^2 \left( w_{xx}^2 + w_{xx}^2 w_x^2 \right) \right] dV + \iiint_{V_c} \frac{1}{2} E_c \left[ z^2 \left( w_{xx}^2 + w_{xx}^2 w_x^2 \right) \right] dV + \iiint_{V_p} \frac{1}{2} E_p \left[ z^2 \left( w_{xx}^2 + w_{xx}^2 w_x^2 \right) \right] dV +$$

$$\iiint_{V_e} \frac{1}{2} E_e \left[ z^2 \left( w_{xx}^2 + w_{xx}^2 w_x^2 \right) \right] dV - \iiint_{V_p} \frac{1}{2} (E_3)^2 \varepsilon_{33}^s dV - \iiint_{V_p} e_{31} E_3 \left[ z \left( w_{xx} + \frac{1}{2} w_{xx} w_x^2 \right) \right] dV -$$

$$\frac{1}{2} k_1 w^2 \left( L + d_{mc}, t \right) - \frac{1}{4} k_3 w^4 \left( L + d_{mc}, t \right)$$

$$\tag{6.9}$$

The $E$ indicates the Young's modulus, where the subscripts are those defined for the respective layers.

The external work is composed of two parts: mechanical work due to the external base acceleration $a_b$ and electrical work caused by the electrical field inside the piezoelectric layers.

$$W = \iiint_{V_s} \rho_s a_b w dV + \iiint_{V_c} \rho_c a_b w dV + \iiint_{V_p} \rho_p a_b w dV + \iiint_{V_e} \rho_e a_b w dV + M_0 a_b w(L + d_c) - \sum_{q=1}^{2} Q_q \cdot v_q \tag{6.10}$$

The $Q_q$ is the charge in piezoelectric layer $q$, $v_q$ represents the voltage at the piezoelectric electrodes. Since there are two piezoelectric layers, $q$ ranges from 1 to 2 in this study.

Although there is discontinuity along the beam axis where the piezoelectric layers end, only the fundamental out-of-plane bending mode is dominate in the responses analysis as justified in Section2.2. This motivates the use of the Ritz method in the process to obtain the Euler-Lagrange equations of motion. Trial functions in the assumed solution by the

Ritz method ensure that the essential boundary conditions are satisfied for convergence of the approximate solution to the accurate result, including the discontinuity [166].

Assuming that the harmonically forced mechanical and electrical behaviors are respectively separable in space and time, the system dynamics may be approximated as a linear combination of linearly independent trial functions and generalized coordinates. For the mechanical responses, it is assumed that

$$w(x,t) = \boldsymbol{\psi}(x)\mathbf{r}(t) = \begin{bmatrix} \psi_1(x) & \cdots & \psi_M(x) \end{bmatrix} \begin{bmatrix} r_1(t) \\ \vdots \\ r_M(t) \end{bmatrix} = \sum_{i=1}^{M} \psi_i r_i \tag{6.11}$$

where $M$ is the total number of trial functions assumed in the summation (which is 3 in the following model development), and $r_i(t)$ is the unknown generalized coordinate. The trial functions $\psi_i(x)$ are approximated by the normal modes of a clamped beam with tip mass given in Eq. (6.12).

$$\psi_i(x) = \cos\frac{\lambda_i}{L}x - \cosh\frac{\lambda_i}{L}x + \zeta_i\left(\sin\frac{\lambda_i}{L}x - \sinh\frac{\lambda_i}{L}x\right) \tag{6.12}$$

$$\zeta_i = \frac{\sin\lambda_i - \sinh\lambda_i + \lambda_i\dfrac{M_0}{M_b}(\cos\lambda_i - \cosh\lambda_i)}{\cos\lambda_i + \cosh\lambda_i - \lambda_i\dfrac{M_0}{M_b}(\sin\lambda_i - \sinh\lambda_i)} \tag{6.13}$$

The $M_b$ is the mass of the beam, $\lambda_i$ is related to the natural frequency of the system, which can be solved by the transcendental equation (6.14).

$$1 + \cos\lambda\cosh\lambda + \lambda\frac{M_0}{M_b}(\cos\lambda\sinh\lambda - \sin\lambda\cosh\lambda) = 0 \tag{6.14}$$

The electrical potential is assumed to vary linearly through the thickness of a piezoelectric layer and become zero-valued outside the piezoelectric domains [32] [164]. The normalized voltage at the electrodes existing at $z = h_s + h_p + h_c$ and $z = -\left(h_s + h_p + h_c\right)$ are assumed to be 1 and -1, respectively, given the mirrored positions of the electrodes considering the laminate sequence about the middle plane of the beam. Consequently, the normalized linear distribution of electrical potential through the thickness of the piezoelectric layers is given by Eqs. (6.16) and (6.17). The Ritz expansion of the electrical potential for the laminated beam is therefore given in Eq. (6.15). The signs of the trial functions of (6.16) and (6.17) are opposite since the locations of the piezoelectric layers are mirrored about the middle plane of the laminated beam.

$$\varphi(z,t) = \boldsymbol{\Phi}(z)\mathbf{v}(t) = \begin{bmatrix} \Phi_1(z) & \Phi_2(z) \end{bmatrix} \begin{bmatrix} v_1(t) \\ v_2(t) \end{bmatrix} = \sum_{q=1}^{2} \Phi_q v_q \tag{6.15}$$

$$\Phi_1(z) = \begin{cases} \dfrac{z - h_s - h_c}{h_p}; & h_s + h_c \leq z \leq h_s + h_p + h_c \\ 0 & elsewhere \end{cases} \tag{6.16}$$

$$\Phi_2(z) = \begin{cases} -\dfrac{z + h_s + h_c}{h_p}; & -\left(h_s + h_p + h_c\right) \leq z \leq -h_s - h_c \\ 0 & elsewhere \end{cases} \tag{6.17}$$

Here $v_q$ represents the voltage at the electrode of the piezoelectric layer $q$.

Substituting Eqs. (6.2), (6.11), and (6.15) into the Eqs. (6.5), (6.9), and (6.10), and then applying the Euler-Lagrange equation (6.18) with respect to the mechanical or electrical generalized coordinates, which are represented by $\xi_m$, the governing equations for the nonlinear energy harvesting system are obtained as shown in Eq. (6.19). Here, Rayleigh proportional damping is assumed and nonlinear coupling between the mechanical and electrical dynamics are neglected.

$$\frac{\partial}{\partial t}\left[\frac{\partial\left(T-U+W\right)}{\partial\dot{\xi}_m}\right]-\frac{\partial\left(T-U+W\right)}{\partial\xi_m}=0 \tag{6.18}$$

$$[M]\ddot{\mathbf{r}}+[D]\dot{\mathbf{r}}+[K]\mathbf{r}+[K_1]\mathbf{r}+\mathbf{F}_{\mathrm{NL}}+[\Theta]\mathbf{v}=\mathbf{f}_m \tag{6.19a}$$

$$-[\Theta]\dot{\mathbf{r}}+\left[C_p\right]\dot{\mathbf{v}}+\left[i_p\right]=0 \tag{6.19b}$$

The nonlinear term is given in (6.20) and $[i_p]$ is the current through the piezoelectric layer.

$$\left[\mathbf{F}_{\mathrm{NL}}\right]_m=\sum_{i=1}^{M}\sum_{j=1}^{M}\sum_{k=1}^{M}\left[K_3\right]_{mijk}r_{mi}r_{mj}r_{mk} \tag{6.20}$$

The matrices in Eq. (6.19) are listed in Eq. (6.21).

$$\begin{aligned}
[M]_{mi}=&\iiint_{V_s}\rho_s\psi_m\psi_i dV+\iiint_{V_c}\rho_c\psi_m\psi_i dV+\iiint_{V_p}\rho_p\psi_m\psi_i dV+\iiint_{V_e}\rho_e\psi_m\psi_i dV+\\
&M_0\left(\psi_m\big|_{x=L}+d_c\frac{\partial\psi_m}{\partial x}\bigg|_{x=L}\right)\left(\psi_i\big|_{x=L}+d_c\frac{\partial\psi_i}{\partial x}\bigg|_{x=L}\right)
\end{aligned} \tag{6.21a}$$

$$\begin{aligned}
[K]_{mi}=&\iiint_{V_s}E_s z^2\left[\frac{d^2\psi_m}{dx^2}\frac{d^2\psi_i}{dx^2}\right]dV+\iiint_{V_c}E_c z^2\left[\frac{d^2\psi_m}{dx^2}\frac{d^2\psi_i}{dx^2}\right]dV+\iiint_{V_p}E_p z^2\left[\frac{d^2\psi_m}{dx^2}\frac{d^2\psi_i}{dx^2}\right]dV+\\
&\iiint_{V_e}E_e z^2\left[\frac{d^2\psi_m}{dx^2}\frac{d^2\psi_i}{dx^2}\right]dV
\end{aligned} \tag{6.21b}$$

$$\begin{aligned}
[K_3]_{mijk}=&\iiint_{V_s}E_s z^2\left(\frac{d^2\psi_i}{dx^2}\frac{d\psi_j}{dx}\frac{d\psi_k}{dx}\frac{d^2\psi_m}{dx^2}+\frac{d^2\psi_i}{dx^2}\frac{d^2\psi_j}{dx^2}\frac{d\psi_k}{dx}\frac{d\psi_m}{dx}\right)dV+\\
&\iiint_{V_c}E_c z^2\left(\frac{d^2\psi_i}{dx^2}\frac{d\psi_j}{dx}\frac{d\psi_k}{dx}\frac{d^2\psi_m}{dx^2}+\frac{d^2\psi_i}{dx^2}\frac{d^2\psi_j}{dx^2}\frac{d\psi_k}{dx}\frac{d\psi_m}{dx}\right)dV+\\
&\iiint_{V_p}E_p z^2\left(\frac{d^2\psi_i}{dx^2}\frac{d\psi_j}{dx}\frac{d\psi_k}{dx}\frac{d^2\psi_m}{dx^2}+\frac{d^2\psi_i}{dx^2}\frac{d^2\psi_j}{dx^2}\frac{d\psi_k}{dx}\frac{d\psi_m}{dx}\right)dV+\\
&\iiint_{V_e}E_e z^2\left(\frac{d^2\psi_i}{dx^2}\frac{d\psi_j}{dx}\frac{d\psi_k}{dx}\frac{d^2\psi_m}{dx^2}+\frac{d^2\psi_i}{dx^2}\frac{d^2\psi_j}{dx^2}\frac{d\psi_k}{dx}\frac{d\psi_m}{dx}\right)dV-\\
&k_3\left(\psi_i\big|_{x=L}+d_{mc}\frac{\partial\psi_i}{\partial x}\bigg|_{x=L}\right)\left(\psi_j\big|_{x=L}+d_{mc}\frac{\partial\psi_j}{\partial x}\bigg|_{x=L}\right)\left(\psi_k\big|_{x=L}+d_{mc}\frac{\partial\psi_k}{\partial x}\bigg|_{x=L}\right)\left(\psi_m\big|_{x=L}+d_{mc}\frac{\partial\psi_m}{\partial x}\bigg|_{x=L}\right)
\end{aligned}$$

$$\tag{6.21c}$$

$$[D]_{mi} = \alpha [M]_{mi} + \beta [K]_{mi} \tag{6.21d}$$

$$[K_1]_{mi} = -k_1 \left( \psi_i \big|_{x=L} + d_{mc} \frac{\partial \psi_i}{\partial x} \bigg|_{x=L} \right) \left( \psi_m \big|_{x=L} + d_{mc} \frac{\partial \psi_m}{\partial x} \bigg|_{x=L} \right) \tag{6.21e}$$

$$[\Theta]_{mq} = -\iiint_{V_p} e_{31} z \frac{d\Phi_q}{dz} \frac{d^2\psi_m}{dx^2} dV \tag{6.21f}$$

$$[\mathbf{f}_m]_m = \iiint_{V_s} \rho_s a_b \psi_m dV + \iiint_{V_c} \rho_c a_b \psi_m dV + \iiint_{V_p} \rho_p a_b \psi_m dV + \iiint_{V_e} \rho_e a_b \psi_m dV + M_0 a_b \left( \psi_m \big|_{x=L} + d_c \frac{\partial \psi_m}{\partial x} \bigg|_{x=L} \right) \tag{6.21g}$$

$$[C_p]_{nq} = \varepsilon_3^s \iiint_{V_p} \frac{d\Phi_q}{dz} \frac{d\Phi_n}{dz} dV \tag{6.21h}$$

From the governing equations (6.19), the terms $[K_1]$ and $\mathbf{F}_{\mathrm{NL}}$ primarily result from the direct influence of the repulsive magnets. By changing the magnet gap $d_2$, the system may take on a monostable or bistable configuration. For a monostable configuration, there is one statically stable equilibrium. According to the influence of the magnet gap on the matrix $[K_1]$, the resonance frequency reduces with decrease of the magnet gap. This trend continues to a point. For still smaller magnet gap $d_2$, two stable equilibria occur and the harvester becomes bistable. In this way, the dynamic responses that may occur depend on the amplitude and frequency of the base acceleration. These dynamics responses may be snap-through, aperiodic, and intrawell vibration. The differences among these three kinds vibration are described in [8].

### 6.2.2 Solutions to the nonlinear system of governing equations

The two piezoelectric layers are in parallel. As a result, the voltage of each piezoelectric layer satisfies

$$v_1 = v_2 = v. \tag{6.22}$$

The governing equations in Eq. (6.19) further simplify to be

$$[M]\ddot{\mathbf{r}} + [D]\dot{\mathbf{r}} + [K]\mathbf{r} + [K_1]\mathbf{r} + \mathbf{F}_{\mathbf{NL}} + [\Theta_1]v_p = \mathbf{f_m} \tag{6.23a}$$

$$-[\Theta_1]^T \dot{\mathbf{r}} + C_{p1}\dot{v}_p = -i \tag{6.23b}$$

where $[\Theta_1]$ is electromechanical coupling associated with the generalized coordinates, $C_{p1}$ is the internal capacitance of the piezoelectric beam, $i$ and $v_p$ are the current and voltage in the harvesting circuit as shown in Figure 6.1(a).

The base acceleration that drives the energy harvester is assumed to have moderate amplitude such that nonlinear harmonics are weakly induced. Thus, the displacement response frequency of the beam is assumed to coincide with the base acceleration frequency $\omega$ [14] [50]. Therefore, the mechanical response are expressed

$$\mathbf{r}(t) = \mathbf{k}(t) + \mathbf{h}(t)\sin \omega t + \mathbf{g}(t)\cos \omega t \tag{6.24}$$

Principles of harmonic or stochastic linearization are then utilized to linearize the governing equations (6.23) [167] [168] [169]. The linearized governing equations are shown in Eq. (6.25).

$$[M]\ddot{\mathbf{r}} + [D]\dot{\mathbf{r}} + \left([K] + [K_1] + [K_e]^{(k_3)}\right)\mathbf{r} + [\Theta_1]v_p = \mathbf{f_m} \tag{6.25a}$$

$$-[\Theta_1]^T \dot{\mathbf{r}} + C_{p1}\dot{v}_p = -i \tag{6.25b}$$

The equivalent linear stiffness matrix accounting for the nonlinearity is given in (6.26).

$$[K_e]_{mh}^{(k_3)} = \left\langle \frac{\partial [\tilde{\mathbf{F}}_{\mathbf{NL1}}]_m}{\partial r_h} \right\rangle = \sum_{j=1}^{M}\sum_{k=1}^{M}\left([K_3]_{mhjk} + [K_3]_{mjhk} + [K_3]_{mkjh}\right)\left[k_j k_k + \frac{1}{2}\left(h_j h_k + g_j g_k\right)\right] \tag{6.26}$$

The $\langle\ \rangle$ indicates the mathematical expectation operator.

When a rectifier with an RC circuit interfaces with a forced piezoelectric beam, the voltage across the piezoelectric layer electrodes is calculated by (6.27) [14]

$$v = \left[ \frac{-[\Theta_1]^T \mathbf{g}}{C_{p1}\pi} \sin^2\beta + \frac{[\Theta_1]^T \mathbf{h}}{2C_{p1}\pi}(2\beta - \sin 2\beta) \right] \sin\omega t + \left[ \frac{[\Theta_1]^T \mathbf{h}}{C_{p1}\pi} \sin^2\beta + \frac{[\Theta_1]^T \mathbf{g}}{2C_{p1}\pi}(2\beta - \sin 2\beta) \right] \cos\omega t$$

$$(6.27)$$

$$\beta = \arccos\left( \frac{\pi - 2\omega C_{p1}R}{\pi + 2\omega C_{p1}R} \right) \tag{6.28}$$

Substituting Eqs. (6.24) and (6.27) into the governing equations (6.25), the coefficients for the sinusoidal terms in Eq. (6.24) are found from the Eqs. (6.29) and (6.30) .

$$-\omega^2[M]\mathbf{h} - \omega[D]\mathbf{g} + \left([K] + [K_I] + [K_e]^{(k_3)}\right)\mathbf{h} - \frac{\sin^2\beta}{C_p\pi}[\Theta_1][\Theta_1]^T \mathbf{g} + \frac{(2\beta - \sin 2\beta)}{2C_p\pi}[\Theta_1][\Theta_1]^T \mathbf{h} = 0 \tag{6.29}$$

$$-\omega^2[M]\mathbf{g} + \omega[D]\mathbf{h} + \left([K] + [K_I] + [K_e]^{(k_3)}\right)\mathbf{g} + \frac{\sin^2\beta}{C_p\pi}[\Theta_1][\Theta_1]^T \mathbf{h} + \frac{(2\beta - \sin 2\beta)}{2C_p\pi}[\Theta_1][\Theta_1]^T \mathbf{g} = \mathbf{f}_m$$

$$(6.30)$$

By integrating the governing equations (6.25) over one period of the harmonic excitation force $2\pi/\omega$, the equations for the constant terms k in Eq. (6.24) are obtained., as shown in Eq. (6.31).

$$\{[K] + [K_I]\}\mathbf{k} + \kappa^{(k_3)} = 0 \tag{6.31}$$

$$\left[\kappa^{(k_3)}\right]_m = [K_3]_{mijk} k_i k_j k_k + \frac{1}{2}\left([K_3]_{mijk} + [K_3]_{mjik} + [K_3]_{mjki}\right) k_i \left[g_j g_k + h_j h_k\right] \tag{6.32}$$

By simultaneously solving the Eqs. (6.29), (6.30), and (6.31), the generalized coordinates are found enabling determination of the physical responses via Eqs. (6.11) and (6.15).

For other nonlinear interface circuits, such as buck-boost converter or synchronized switch harvesting on inductor electronic interface, with an expression of the output voltage $v_p$ across the piezoelectric beam determined through the respective derivation [50] [26] [27], the procedures described above may be applied for the responses analysis.

## 6.3 Numerical verification and comparison to analysis

In order to verify the accuracy of the approximate analytical solution to the nonlinear system of governing equations, fourth-order Runge-Kutta numerical integration is undertaken for the equation system (6.23). Due to the potential for multiple steady-state dynamic behaviors, 10 separate simulations are taken using normally distributed and randomly selected initial conditions (i.e. displacements, velocities, and voltages) for each combination of base acceleration and system parameters. The simulation duration is set to be 400 periods of the harmonic excitation to ensure steady-state conditions develop. The parameters of the system are given in Tables 6.1 and 6.2. The beam length $L$ and gap between repulsive magnets $d_2$ are taken as the outcomes of the optimization, discussed in detail in Section 6.4, but it is sufficient at this stage to use these key parameters for contrast to the simulation. Cubic magnets with side lengths of 6.35 [mm] are used to introduce nonlinearity. The mass of the magnet and mass of the holder contribute to the total mass at the cantilever tip, which is 15 g. Table 6.2 provides the properties for layers of the laminate beam. In addition to the nominally optimal parameter combination of harvester beam design studied, additional parameter combinations are examined here to comprehensively verify the analytical model. These four cases involve a combination of 7% increase or decrease in beam length $L$ with 5% increase or decrease in magnet gap $d_2$.

Table 6.1 Parameters for system design and excitation conditions.

| $L$ (mm) | $b_0, b_L$ (mm) | $h_s$ (mm) | $h_p$ (mm) | $h_z$ (mm) | $d_1$ (mm) | $d_c$ (mm) | $d_{mc}$ (mm) | |
|---|---|---|---|---|---|---|---|---|
| 37.95 | 32 | 0.04 | 0.19 | 0.03 | 10 | 6.58 | 8.9 | |
| $d_2$ (mm) | $M_0$ (g) | $M_a$ (MA/m) | $V_{mag}$ (cm$^3$) | $a_b$ (m/s$^2$) | $L_p$ (mm) | $b_p$ (mm) | $R$ (k$\Omega$) | $C_L$ ($\mu$F) |
| 10.2 | 15 | 1.61 | 0.768 | 5 | 27.8 | 18 | 100 | 10 |

Table 6.2 Material properties.

| $E_s$ (GPa) | $E_p$ (GPa) | $E_c$ (GPa) | $E_e$ (GPa) | $\rho_s$ (kg/m$^3$) | $\rho_p$ (kg/m$^3$) | $\rho_c$ (kg/m$^3$) | $\rho_e$ (kg/m$^3$) | $\varepsilon_{33}$ (nF/m) | $e_{31}$ (C/m$^2$) |
|---|---|---|---|---|---|---|---|---|---|
| 26 | 60.6 | 128 | 3.5 | 1900 | 7800 | 8940 | 1540 | 13.16 | -16.6 |

Given the 5 m/s$^2$ amplitude harmonic base acceleration identified in Table 6.1, Figure 6.2 shows the piezoelectric cantilever responses in terms of (a) displacement amplitude at the beam tip, (b) rectified output voltage, (c) strain at the fixed end on the bottom beam surface, and (d) static equilibrium position. Analytical results are shown as curves whereas simulation results are the open markers. From Figure 6.2(d), when the magnet gap decreases (case 5) or beam length increases (case 4) from the nominal case 1 design parameters, non-zero static equilibria appear. This indicates that for case 4 and case 5 an increase in beam length and a decrease in magnet gap result in a bistable energy harvester configuration. At higher frequencies of base acceleration, the residual strain caused by the non-zero static equilibria contributes to the greater total strain shown in Figure 6.2(c), despite exhibiting small amplitudes of displacement, Figure 6.2(a).

The Table 6.3 quantitatively compares the analytical and simulation results from Figure 6.2. The frequency bandwidth shown in Table 6.3 is determined by the frequency range in the harmonic voltage responses over which the voltage amplitude is greater than or equal to one-half of the maximum computed voltage amplitude. The corresponding root mean square (RMS) values are calculated inside this frequency bandwidth.

Table 6.3 Analytical and simulation results comparison.

| | Case 1 | | Case 2 | | Case 3 | |
|---|---|---|---|---|---|---|
| | Anal | Sim | Anal | Sim | Anal | Sim |
| Bandwidth [Hz] | 11.86 | 12.08 | 9.46 | 9.52 | 11.74 | 11.79 |
| RMS value of rectified voltage [V] | 13.87 | 10.7 | 13.87 | 11.52 | 12.43 | 10.52 |
| RMS value of strain[$\mu\varepsilon$] | 420.84 | 427.65 | 376.27 | 378.11 | 332.62 | 333.9 |

As shown in Table 6.3, the differences between the analytical and simulation results for the frequency bandwidth of large amplitude dynamics and RMS strain at the clamped end are less than 2%. Based on the derivation in Section 6.2, only the fundamental harmonic term is employed to approximate the system responses, thus neglecting diffusion of energy to higher order harmonics, which contributes to the overprediction of the rectified RMS voltage by the analysis as seen in Table 6.3. Despite this discrepancy, from Figure 6.2(b) the voltage responses of the first three cases show good agreement for the influence of 5%

decrease in magnet gap and 7% decrease in beam length. On the other hand, for the bistable

cases (case 4 and 5) in Figure 6.2(a), the harmonic response and steady state assumptions

of analysis lead to less agreement with simulation due to the chaotic responses found by

numerical simulation that are unable to be reproduced by the harmonic analysis. At higher

frequencies, Figure 6.2 uniformly reveals good quantitative agreement between simulation

and analysis, where the significance of nonlinear response is relatively low.

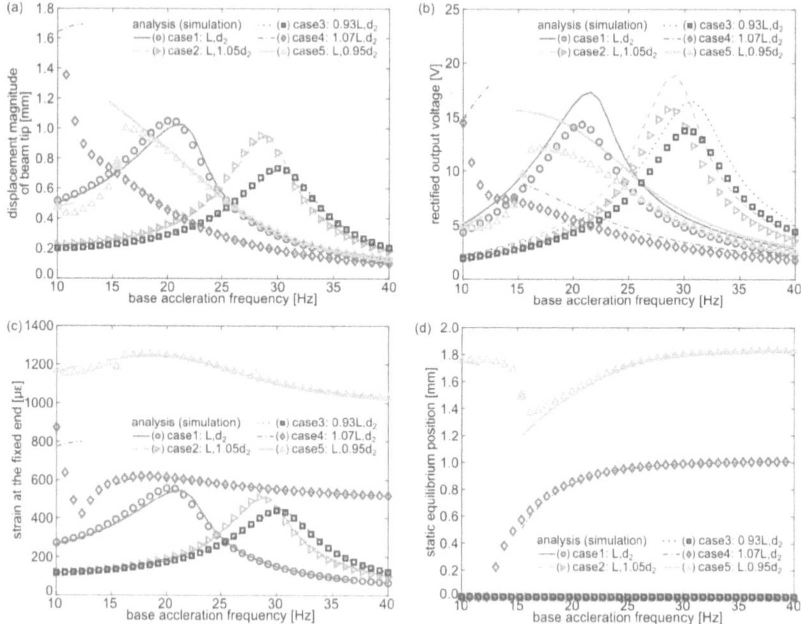

Figure 6.2 Frequency responses of (a) beam tip displacement amplitude, (b) rectified output

voltage across the resistor, (c) strain of the beam at the fixed end, and (d) static equilibrium

position.

159

Overall, comparing with the simulation results, the analytical model captures the main characteristics of the nonlinear energy harvesting system for both monostable and bistable configurations resulting from the key parameter shifts of beam length and magnet gap.

## 6.4 Optimization for a robust energy harvester system via genetic algorithm

Multiple attributes of an energy harvesting system contribute to a robust system design for persistent DC power delivery. To understand how to design parameters distinctly participate to yield exceptional overall performance for the laminated piezoelectric cantilever, here a multi-objective, genetic algorithm (GA) optimization method is established and then utilized to scrutinize what combinations of design elements give rise to optimality.

### 6.4.1 Multi-objective optimization constraints and objectives

To consider a practical design problem, several overall constraints are set in the exploration of feasible design parameters. The space to install energy harvesters in application is likely confined [170]. The maximum length $L_{\max}$, width $b_{\max}$, and thickness $t_{\max}$ of the maximum operating volume are set to be 70 mm, 40 mm, and 4 mm, respectively. The thickness extent is the most limited since the beam vibrates in this axis. To generally minimize dynamic strain and prolong working life, the peak amplitude of displacement must be limited. Specifically, the global constraints based on these limitations for the optimization are as follow:

1. The sum of the beam length $L$, holder extension $d_1$, and magnet gap $d_2$ must be less than maximum length $L_{max}$.

2.  The beam length must be greater than 30 mm and less than 60 mm to warrant assumptions of vibration like a beam in contrast to a plate.

3.  The displacement amplitude at the beam tip must be less than $t_{max}/2$ to not exceed the maximum permitted thickness of the operating volume.

4.  The magnet gap $d_2$ must be less than 20 mm and greater than 5 mm, to minimum unnecessary cases of extreme nonlinearity.

5.  The displacement amplitude at the beam tip must be greater than 1 mm, otherwise GA searches use excessive time for irrelevant and poor-performing combinations of design parameters.

The following three objectives are utilized for the multi-objective optimization considered in this report.

1.  Frequency range $f_{range}$. The $f_{range}$ is calculated from the voltage amplitude based on the concept of frequency bandwidth defined in Section 6.3. For a bistable configuration of the energy harvester, only the snap-through response is considered to determine the frequency range.

2.  Total RMS of rectified output voltage $V_{rms}$ in $f_{range}$. For a monostable system, the values inside the frequency range $f_{range}$ are included in the calculation. For a bistable system, only the values associated with snap-through inside the frequency range contribute to the cost function of $V_{rms}$.

3.  Total RMS of strain at the clamped beam end on the bottom surface $S_{rms}$ in $f_{range}$. The calculation is similar to the procedures described in the objective 2. The

difference occurs when the system becomes bistable. Because the existence of non-zero static equilibrium positions will cause residual strain in the whole frequency range. The RMS strain for bistable configurations include the contributions from the residual strain and the dynamic strain amplitude.

The multi-objective optimization is formally defined by a weighted sum of the individual objectives [171]. A nondimensionalization is applied before the summation of components. The linear response for a piezoelectric cantilever is adopted to obtain the nondimensionalization constants. Since the bandwidth for a linear response is narrow when compared to the nonlinear dynamic behavior, the resonance frequency $f_0$ of the linear piezoelectric cantilever is used to normalize the $f_{range}$. The voltage $V_0$ at resonance for the linear piezoelectric cantilever is used to normalize $V_{rms}$. Finally, the strain $S_0$ at the fixed end on the bottom surface of the linear piezoelectric cantilever at resonance is used to normalize $S_{rms}$. These values are found to be

$$f_0 = 25\,Hz,\ V_0 = 65\,V,\ S_0 = 1365\,\mu\varepsilon \tag{6.33}$$

The total objective employed for optimization is

$$\cos t = -w_f \frac{f_{range}}{f_0} - w_v \frac{V_{rms}}{V_0} + w_s \frac{S_{rms}}{S_0} \tag{6.34}$$

$$w_f + w_v + w_s = 1 \tag{6.35}$$

The $w_f$, $w_v$, and $w_s$ are respectively weights for the objectives of frequency range, rectified output voltage, and strain condition. Since an energy harvester can deliver high output power in a wide frequency range with a reasonable low strain condition is preferred, with the total objective defined in Eq. (6.34), the smallest value of the total objective indicates the optimal design.

The genetic algorithm (GA) optimization developed for this research is inspired through the specific strategies described in Haupt and Haupt [171]. A total of 100 generations including populations with 80 individuals are used for broad evaluation of the parameter space. Each individual is a combination of the design parameters defined for the several optimization studies considered in the following sections. Each individual is evaluated for the three distinct objectives: frequency range, RMS voltage, and RMS strain. The cost function for each individual is then computed by Eq. (6.34) according to the multi-objective weighting. The initial population of 80 individuals is generated through random selection inside the constraint range of each design parameter. After evaluating and ranking each individual in the population on the basis of cost function value, the best 50% of the population is selected as parents to generate 40 offspring for the next generation. The remaining 50% of the next generation is the best 50% of the prior population. A mutation rate of 20% is also included to randomly mutate 20% of the next generation to ensure the diversity inside each population.

In the following sections unless otherwise indicated, the parameters given in Tables 6.1 and 6.2 are employed. The base acceleration frequency range is from 10 Hz to 40 Hz with a constant amplitude 5 m/s$^2$, which is characteristic of ambient vibrations in heavy industry or automotive applications. Three optimization studies are undertaken to uncover the origins of optimal robust DC power delivery from the laminated piezoelectric cantilever to a fixed load condition. The detailed influence of the resistance on the performance of the piezoelectric energy harvester may be found in [68] [50].

### 6.4.2 Optimization results and discussion

*6.4.2.1 Optimization 1: beam length $L$ and magnet gap $d_2$*

A first optimization problem with the design parameters of beam length $L$ and distance between repulsive magnets $d_2$ is considered to understand how the nonlinear forces associated with magnetic repulsion influence the ideal energy harvesting beam design. When the multi-objective cost function weights $w_f$, $w_v$, and $w_s$ are the same value (1/3), a system with beam length $L$=37.95 mm and magnet gap $d_2$=10.17 mm is identified as optimal. This parameter combination is shown as the star marker in Figure 6.3. Figure 6.3 presents the values of the cost function as the shading for a wide range of combinations of beam length and magnet gap. Region I conflicts with constraint 1. Regions II and V are neglected because they violate constraint 5. For harvesters designs in region II, with the decrease of the magnet gap, more energy is demanded to pass the potential barrier to realize snap-through behavior. Therefore, in region II only small amplitude of intrawell behaviors are possible. Because of the residual stress caused by the non-zero static equilibria, the intrawell responses result in large strain and low output voltage. For designs inside region V, the resonance frequency of the system lies sufficiently outside the frequency range of interest in this research, which results in low rectified output voltage flow. In contrast, region III includes the designs with displacement amplitude larger than 2 mm, which compromises longevity of the harvester and conflicts with the volumetric space constraints.

The star label resides in the region IV, where all combinations of the magnet distance and beam length satisfy the optimization constraints.

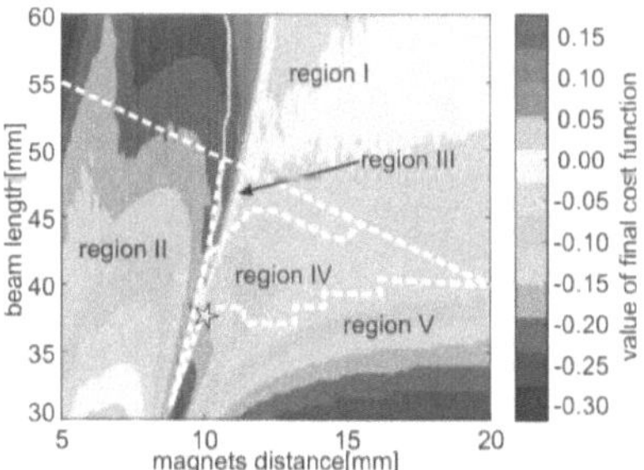


Figure 6.3 Cost function values of different combination of beam length and magnet gap.

The results of this optimization may be assessed in relation to the findings from the numerical verification of the analytical model in Figure 6.2. The case 1 in Figure 6.2 provided the optimal result obtained in this Section 6.2.1 optimization when all weights are the same. For cases 4 and 5, the length of the laminated cantilever is increased or the magnet gap is decreased, enhancing the nonlinear interaction between repulsive magnets and causing bistability. The bistability is seen to result in high strain at the clamped end of the cantilever Figure 6.2(c), which practically reduces the working life of the harvester due to the susceptibility to early failure.

The optimal design case 1 and the sub-optimal designs of cases 2 and 3 are monostable, such that the resting equilibria have zero displacement of the beam tip. The findings of Figure 6.2 thus show that a decrease in beam length or increase in magnet gap is desirable

165

to sustain a monostable, and therefore long-life platform design, if the optimal design
parameters are unable to be perfectly achieved.

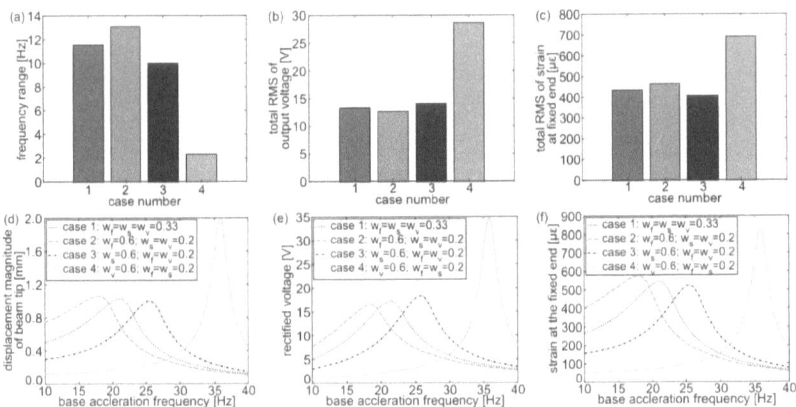

Figure 6.4 Cost function values of (a) frequency range, (b) total RMS of the rectified output
voltage inside the frequency range, and (c) total RMS of the strain at the fixed end on the
bottom surface inside the frequency range. Corresponding responses of (d) displacement
amplitude of the beam tip, (e) rectified output voltage, and (f) strain at the fixed end on the
bottom surface.

In order to study the influence of weights in this optimization problem, three more
optimization problems are considered. Here, the weights assigned to compute the cost
function values are differed. Case 2 corresponds to the case that $w_f$, $w_v$, and $w_s$ are
respectively 0.6, 0.2, and 0.2. Case 3 corresponds to the case that $w_f$, $w_v$, and $w_s$ are
respectively 0.2, 0.2, and 0.6 for $w_f$, $w_v$, and $w_s$. Finally, case 4 corresponds to the case

that $w_f$, $w_v$, and $w_s$ are respectively 0.2, 0.6, and 0.2. These unique cases make the individual cost function objectives relatively dominant: (case 2) frequency range $f_{range}$, (case 3) strain condition $S_{rms}$, and (case 4) rectified voltage $V_{rms}$.

The optimal designs identified for the three cases are respectively $L$=37.37 mm, $d_2$=9.94 mm for case 2, $L$=38.38 mm, $d_2$=10.51 mm for case 3, and $L$=43.86 mm, $d_2$=16.13 mm for case 4. Figures 6.4(a,b,c) present the cost function values computed for harvesters optimized with the different weighted cases. Figures 6.4(d,e,f) respectively show the responses of beam tip displacement, rectified voltage, and strain at the cantilever end of four cases. Since the magnet gap is more influential in introducing nonlinearity, when the frequency range is dominant (case 2) in the optimization, a smaller magnet gap is optimized to introduce greater nonlinearity that leads to a wider frequency range. Comparatively, when the strain condition is dominant (case 3), the magnet gap increases to realize the nonlinear resonant behaviors at a comparatively higher frequency as shown in Figure 6.4(d). The responses of the cases 1, 2, and 3 in Figures 6.4(e,f) indicate an increase in output voltage at high frequencies of base acceleration due to a shifting of the nonlinear resonance to higher frequencies even with a smaller strain at the cantilever end. This explains why the optimal magnet gap in case 3 is greater than those for cases 1 and 2. For a similar reason, when the voltage is dominant (case 4), the magnet gap is further increased with a much longer beam length to highly increase the rectified output voltage by increased resonant frequency. In comparison with the other three cases, the case 4 optimization demonstrates a tremendous drop in frequency range and increase in strain and output voltage in Figures 6.4(a,b,c). From the responses shown in Figure 6.4(d), the optimal design in case 4 is more similar to a linear configuration for its narrow bandwidth, which illustrates the sensitivity of the voltage objective to the weights change.

### 6.4.2.2 Optimization 2: beam length $L$, magnet gap $d_2$, and beam width at free end $b_L$

Multiple investigations have concluded that a trapezoidal shape for a piezoelectric cantilever results in more uniform strain distribution and often increase in output power [32] [146]. Therefore, the beam width at the free end $b_L$ of the laminated piezoelectric cantilever is introduced as an additional design variable for the optimization. The additional constraint for $b_L$ is that it may vary from zero to the maximum width value, which is 40 mm as stated in Section 6.1. Four optimization cases, having the same variation of objective weights employed in Section 6.2.1, are considered here. As in Section 6.2.1, these cases are also referred to as cases 1, 2, 3, and 4 based on the weights used in the optimization. The optimal designs obtained for the four cases including the beam length, beam width, and magnet gap are schematically shown in Figure 6.5(a). The black dotted curve overlaid on each beam shape schematic is the outline of the piezoelectric layer. The grey squares represent the positions of the magnets, and the white rectangles are the magnet holders. The clamp location for the beams corresponds to the bottom of each respective schematic, so that the holder and magnet are installed at the opposite end. The repulsive magnets are positioned opposite of the magnet holder and the magnet gap is identified in the Figure 6.5 according to the spacing shown. The mass labeled is the total tip mass, which is the combination of the magnet and the holder mass. The strain distribution of the cantilever on the bottom surface is also shown as the shaded contour in Figure 6.5. The magnitude of the strain is nondimensionalized by the strain at the fixed end. The average values of the nondimensional strain $\varepsilon_{avg}$ are reported on the surfaces and correspond to the average nondimensional strain over the area covered by the PZT-5H layer.

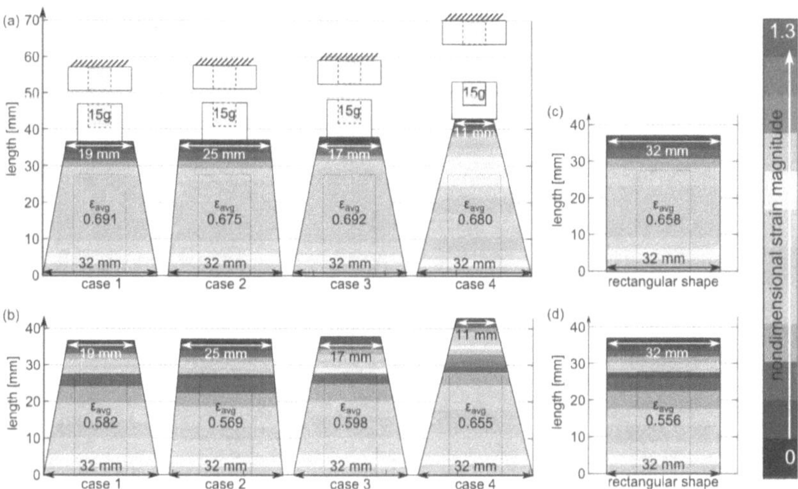



Figure 6.5 (a) Strain contour plot from analysis and (b) strain contour plot from finite element simulation of optimal designs. (c) Strain contour plot from analysis and (d) strain contour plot from finite element simulation of a rectangular shape for comparison.

Figure 6.5(a) presents the analytical reconstruction of the strain distributions whereas Figure 6.5(b) presents strain results obtained from finite element simulations (ABAQUS) of the piezoelectric cantilevers having the same optimized geometries, layering, and material properties. The schematics of the magnets in Figure 6.5(b) are omitted for sake of brevity. Since in the analysis the strain is assumed to be uniform through the beam width and equal to the axial strain, the finite element results are also presented with the axial strain mapped equally over the beam width. Here, both the analytical and finite element strain distribution for a rectangular beam shape are respectively provided in Figures 6.5(c,d) for comparison. The dimensions of the rectangular beam include a constant beam width of 32 mm along the beam axis.

Comparing the finite element and analytical results in Figures 6.5(a,b), the strain distribution around the end of the PZT cannot be exactly captured in the analysis due to the use of a limited number of trial functions. This is especially evident for case 4 in Figure 6.5(a), where the finite element simulation Figure 6.5(b) predicts a greater nondimensional strain at the position of the discontinuity than that suggested by the analysis in Figure 6.5(a). Despite the relatively minor discrepancies, both analytical and finite element results demonstrate that a more uniform strain distribution is achieved over the tapered beam shape. The analysis and simulations both agree that the tapered beam strain distributions, Figures 6.5(a,b), are more uniform when compared to the corresponding rectangular beams, Figures 6.5(c,d). A greater average nondimensional strain over the piezoelectric layers indicates that the tapered beams deliver comparatively higher output electrical power to the loads than the rectangular beams. Therefore, considering the four cases of optimization, whether all objectives are equally weighted (case 1) or the individual objectives are set to be dominant (cases 2, 3, and 4), the optimization results show that tapered beams with strategic beam length and magnet gap deliver improved global performance. When the strain condition (case 3) or rectified voltage (case 4) objective is dominant in the optimization, a higher taper ratio defined as $b_L / b_0$ is optimized. This further supports the conclusion that tapered beam shapes deliver higher power by the more uniform strain distribution, even though the piezoelectric layers only cover part of the beam. In addition, comparing with the optimal designs of cases 1 and 2, the magnet gap increases for the cases 3 and 4. This is especially evident for case 4, where the magnet gap is greatly increased to reduce nonlinearity and promote more linear response behavior. This agrees with the conclusion in Section 6.2.1.

### *6.4.2.3 Optimization 3: beam length $L$, magnet gap $d_2$, and tip mass $M_0$*

Since applications often envision collocation of the harvester with the powered microelectronics, it is desirable to limit the harvester mass to lead to a compact, lightweight energy harvesting platform for system integration. In the third optimization study, the beam length, magnet gap, and tip mass are taken as design parameters to optimize. Here a maximum tip mass 25 g is taken as additional constraint. Four cases with different weights setting as stated in Section 6.2.1 are studied.

The optimal designs for the four cases are schematically shown in Figure 6.6. As described in Section 6.2.2, the light grey square and the white rectangle represent the magnet and the holder. The mass labels are the total magnet and holder mass. Since the extended length of the magnet holder at beam free end are constant, the shape or the material of the holder may change to ensure the optimal tip mass. In the schematics, the holder shape is changed to illustrate a change of optimized tip mass. The additional characteristics schematically shown in Figure 6.6 are similar to those in Figure 6.5. As first observed through the optimization of Section 6.2.1, the magnet gap greatly increases for case 4 to promote linear dynamic behavior and greatest RMS voltage in a narrow frequency range. For the cases 1,

2, and 3 in Figure 6.6, sufficient nonlinearity results from the repulsive magnet interaction to lead to a monostable configuration. Additionally, since the tip mass is directly related to the external force induced at the beam tip, when considering the tip mass as a design variable a larger mass corresponds to a greater tip force that increases the strain along the beam length and correspondingly delivers high output power. Therefore, when emphasizing the minimization of strain by case 3, the optimal design converges to a small tip mass shown in Figure 6.6. In contrast, when the rectified output voltage becomes the dominant factor in the optimization shown by case 4, the maximum allowable tip mass is selected with a short beam length to lead to a higher frequency of resonance, as similarly observed in Section 6.2.1 . For the cases 1 and 2, the optimized tip mass results from the respective balance of the multiple objectives in the cost function evaluations.

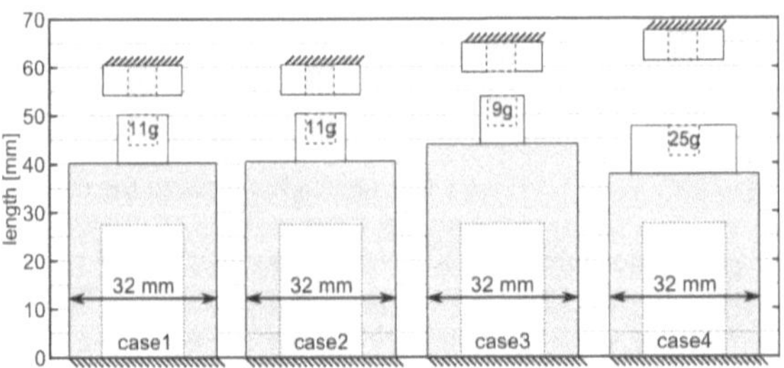


Figure 6.6 Optimal designs for optimization 3.

From the corresponding cost function values shown in Figure 6.7, the change in the tip mass causes up to 27 V change in the RMS voltage, while the RMS strain increases up to 900 $\mu\varepsilon$ for case 4. This confirms the significant influence of the tip mass selection on the output voltage and strain at the cantilever fixed end. In contrast, the influence of tip mass on optimizing the frequency range is weak since the beam length and magnet gap are primarily shifted to promote a broader frequency range of the nonlinear resonant behavior. These results show that it is valuable to include the tip mass as a design variable in optimization to ensure that a combination of harvester design parameters is optimized having overall greatest lifetime and best power delivery.

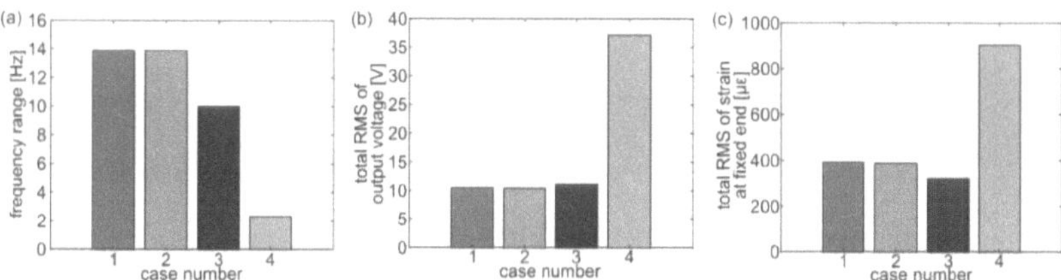

Figure 6.7 Cost function values of four optimal designs. (a) Frequency range. (b) Total RMS of the output voltage. (c) Total RMS of the strain at the fixed end.

From all optimal designs for case 4 in the three optimization problems considered in this Section 6.4, when the voltage objective is dominant in the optimization extreme design conditions are reached. In other words, case 4 results in the maximum permitted beam length in optimization 2 and the maximum tip mass in optimization 3. The frequency responses of the optimal designs for the optimized results for case 4 are similar to linear responses, as seen by the examples in Figure 6.4(d,e,f). Based on the narrowband resonant behavior, such linear systems may easily lose superiority of performance when excitation conditions change. Therefore, in an optimization of an energy harvesting system, a strict

173

optimization on the basis of maximizing output voltage is not recommended. Comparatively, preferring optimizing on the basis of frequency range, strain at the fixed beam end, or the more balanced multi-objective optimization lead to robust energy harvesters that may serve in practical environments.

## 6.5 Experimental system description

The key findings from the analytical model and optimization investigations are then validated through controlled experiments. Based on the optimal designs of the laminated piezoelectric cantilevers determined through case 1-3 in Section 6.2, three beams shown in Figure 6.8(a) are cut from off-the-shelf piezoelectric energy harvesters by Midé Technology. The shapes are cut using a CNC router. Due to limitations of creating the cutout cantilever shapes, the optimal geometries are unable to be exactly achieved although they are closely emulated. The beams 1, 2, and 3 shown in Figure 6.8(a) have experimentally identified parameters given in Table 6.4, while the remaining parameters not given are provided in Table 6.2.

Figure 6.8(b) presents the experimental setup used to examine the three piezoelectric cantilevers. For each experiment, the piezoelectric beam is clamped to an aluminum frame mounted to an electrodynamic shaker table (APS Dynamics 400). At the free end of the piezoelectric beam, an aluminum magnet holder is installed. An opposing magnet holder is affixed to the shaker table at a location immediately adjacent to the beam tip along the beam axis, as shown in Figure 6.8(b). The shaker table is driven by a controller (Vibration Research Controller VR9500) and amplifier (Crown XLS 2500) with an accelerometer (PCB Piezotronics 333B40) to provide control signal feedback to ensure the base acceleration amplitude remains 5 m/s$^2$ at all frequencies examined. Two laser displacement sensors (Micro-Epsilon ILD-1420) measure the absolute displacement of the beam tip and the shaker table. A bridge rectifier (1N4148 diodes) with a smoothing capacitor $C_r$ and resistive load $R$ are connected to the piezoelectric beam to quantify the output power

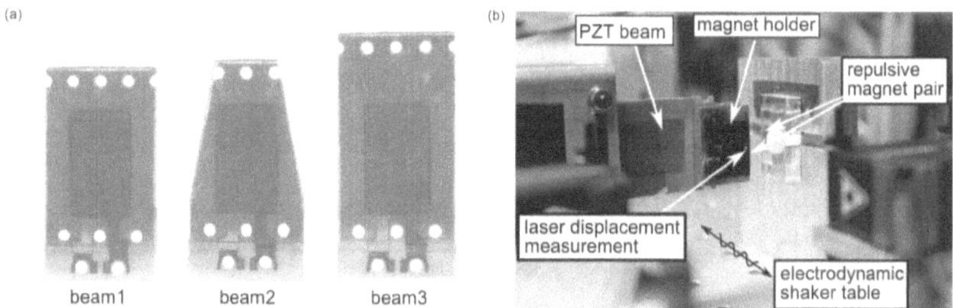


Figure 6.8 (a) Three beams shapes for experiments. (b) Photograph of the experiment platform.

Table 6.4 Parameters for three beams identified from experiments.

| | $L$ (mm) | $b_L$ (mm) | $h_s$ (mm) | $h_c$ (mm) | $d_2$ (mm) Monostable | $d_2$ (mm) Bistable | $\varepsilon_{33}$ (nF/m) | $e_{31}$ (C/m$^2$) | $M_0$ (g) |
|---|---|---|---|---|---|---|---|---|---|
| Beam 1 | 37.4 | 32 | 0.034 | 0.02 | 10.8 | 10.4 | 30.15 | -14.54 | 16.6 |
| Beam 2 | 37.4 | 18 | 0.04 | 0.02 | 10.6 | 10.3 | 30.15 | -14.54 | 16.6 |
| Beam 3 | 45 | 32 | 0.034 | 0.02 | 11.8 | 11.6 | 30.15 | -14.54 | 7.0 |

## 6.6 Experiment validation

The experiments examine both monostable and bistable configurations of beams 1, 2, and 3. Figures 6.9(a,b) respectively show the displacement amplitude of the beam tip and rectified output voltage of the three beams for a monostable configuration. Figures. 6.9(d) and (e) are the corresponding responses for a bistable configuration of each beam. Analytical predictions are shown as the thick curves whereas the experimental results are presented using thin curves in the same respective line style. Because the strain at the clamped ends of the beams cannot be measured without influencing the laminated beam mechanical properties, the strain responses shown in Figures 6.9(c,f) are calculated from finite element analysis (FEA) using ABAQUS. In the FEA, a solid model of the exact beam composition is used, where all parameters employed are provided in Tables 6.2 and 6.4. Rigid elements are included to model the aluminum magnet holder, and a mass element is used to model the beam tip mass. After establishing the finite element model, the displacement amplitudes shown in Figures 6.9(a,d) are applied as boundary conditions for the FEA to compute a static load analysis that provides values of the clamped end strain, using the strain at the middle of the clamp.

From the displacement responses shown in Figures 6.9(a,d), the large displacement amplitudes at the bifurcations are in good agreement between the analysis and experimentation. Since the stiffness of the beam is increased due to the rigid cross-section assumption in Euler-Bernoulli beam theory, the analytical strain predictions in Figures 6.9(c,f) are relatively greater than the FEA results using the experimental data of displacement amplitude. This in turn increases the predicted analytical output voltage in Figures 6.9(b,e) compared to the experimental levels. Electrical component losses and higher order harmonics are also neglected in analysis, which may also contribute to the overprediction of rectified voltage. The grey shaded areas in Figures 6.9(d,e,f) correspond to multi-harmonic or chaotic dynamic behaviors that are not predicted by analysis due to steady-state assumptions in the solution formulation. Despite these discrepancies, the overall quantitative agreement between the experimental and analytical results in Figure 6.9 exemplifies that the model formulation accurately replicates the electromechanical behaviors of the piezoelectric laminated energy harvester.

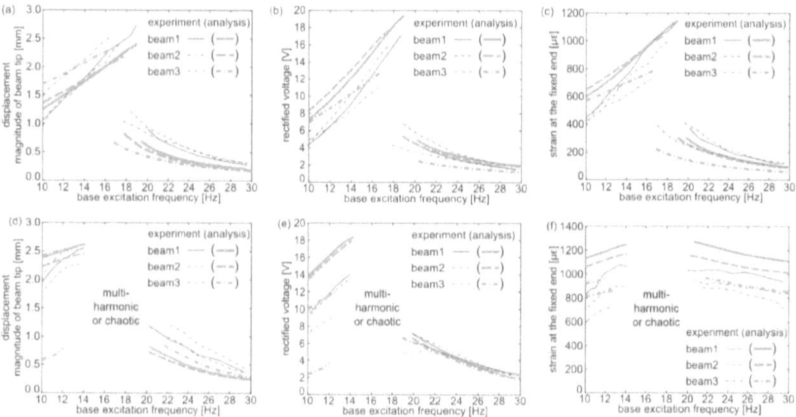

Figure 6.9 Frequency responses for three beams. (a) Displacement amplitude at the beam tip, (b) rectified output voltage, and (c) strain at the fixed end for monostable configurations. (d), (e), and (f) Corresponding responses for bistable configuration.

The results of Figure 6.10 establish the efficacy of the optimization created through this research. The results of monostable cases of all three beams suggest a higher output voltage with a lower strain level and a wider frequency range (red or blue bars) when compared to the bistable configurations of the energy harvesters (green or cyan bars). This emphasizes the global superiority of monostable platforms over bistable system configurations to achieve broadly robust energy harvesting performance and agrees with results obtained throughout Section 6.4. In addition, the tapered beam 2 exhibits quantitatively less strain at the clamped end than the rectangular beam 1 of the same length, Figure 6.10(c), while the beam 2 is within 1 V RMS of the rectified output voltage of beam 1, Figure 6.10(b). This result agrees with the findings of Section 6.4.2.2. Furthermore, when the tip mass decreases to 7 g for beam 3, the frequency range may still be broad by introducing sufficient nonlinearity as shown in Figure 6.10(a). Yet, the output voltage and strain for beam 3 in Figures 6.10(b,c) are decreased comparing with the results of the other two beams, which supports the influences uncovered in Section 6.4.2.3. Thus, considering the design constraints examined here, piezoelectric laminated cantilevers with tapered shapes, monostable nonlinearity, and moderate lengths deliver the most performance robust DC power delivery.

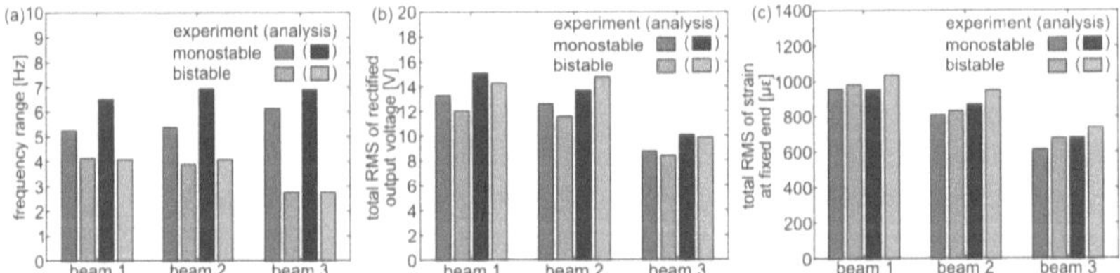

Figure 6.10 Comparison of cost function values between experiment and analysis for three beam designs. (a) Frequency range. (b) Total RMS of the rectified output voltage. (c) Total RMS of the strain at the fixed end.

178

**6.7 Conclusion**

This research provides the first illumination of the coupled influences of nonlinearity and beam shape design on the resulting mechanical integrity and versatile electrical power generation of a nonlinear energy harvesting system. Here, an analytical model is established and verified through numerical simulations and experimental examinations for key influences of nonlinearity, beam shape, and tip mass select result in robust system designs. Multi-objective optimization guides attention to strategic combinations of design parameters based on the specific optimization variables and relative weights given to the objective functions. Trapezoidal beam shapes are found to promote large output voltage without yielding high strain at the clamped end, which promotes practical longevity of the system. Bistable implementations of the nonlinear energy harvesters are found to be undesirable for sake of large residual strain at the clamped end, even though they may deliver wide frequency range of operation and high output voltages. Therefore, both the optimization results using the new analysis and experiment validations indicate that a trapezoidal beam in a monostable configuration is globally preferred.

## Chapter 7 Conclusions

This research studies the energy harvesting system leveraging the nonlinearity of the mechanical and electrical platforms when subjected to complex ambient vibration. To conclude the book, a summary of key discoveries is provided in this chapter. In addition, the directions for future work and possible applications are also discussed.

### 7.1 Dynamic characterizations for nonlinear energy harvesting systems subjected to combined harmonic and stochastic excitation

This research formulated a range of first principles-based models for the first time that directly correlate the mechanical, electrical, and multiphysics domains with the corresponding nonlinear behaviors far from equilibrium resulting from stochastic and deterministic perturbations. The previous studies in the state-of-the-art articulate the dynamics of the energy harvesting system introduced by either pure harmonic or stochastic excitation [9] [20] [21] [41]. In Chapter 2, an analytical approach based on the equivalent linearization method is established to account for contributions to dynamics from harmonic and stochastic vibration inputs. The proposed linearization technique help decouple the harmonic and stochastic contributions, as such superposition is feasible to be applied to examine the dynamic responses. In addition, a weighted Gaussian joint distribution is adopted in the analytical model to approximate the real distribution of the random process resulting from the stochastic inputs for the far-from-equilibrium systems. In terms of the electrical power calculation, the fundamental harmonic component is taken to represent the non-smooth electrical response. With validations from simulations and experiments, the proposed analytical model is verified to be able to describe the unique dynamics of a system considering the multiphysics coupling and nonlinearity when subjected to stochastic and deterministic perturbations. Utilizing the analytical model, the knowledge of the consequences of asymmetry and noise on the dynamic features has been revealed. The findings through the proposed analytical model shed light on the complex dynamics of the integrated systems when incorporating the source stochasticity and harvester imperfection, which may assist in the real-world implementation of the energy harvesting systems for improved, robust power delivery. Beyond vibration energy harvesting, the proposed theoretical tool is also valuable to facilitate insights on future studies where multiphysics combine towards the resulting system dynamics.

**7.2 Nonlinear energy harvesting systems subjected to periodic impulse excitation**

The environment that energy harvesters are likely deployed within may experience complex excitation conditions. Due to the complexity introduced by the nonlinear coupling among the multiphysics domains, numerical approaches are frequently employed to characterize the dynamics under complex perturbations [80] [81] [82], which is inefficient for identifying optimal implementations accounting for the difficulty in deciphering the high-dimensional data from simulations. In Chapter 3, impulse excitation is taken as an example to construct a predictor model by integrating data-driven analysis with the physical insights of a first-principles model for the nonlinear energy harvesting piezoelectro-mechanical system. In the predictor model, supervised learning is selected to map the system parameters to the dynamic mechanisms with mapping functions determined by neural network algorithms. Utilizing the demarcation of parameters in the predictor model, the strategies for maximizing DC power delivery have been probed with impedance analysis. Through the proposed predictor model, the knowledge on the relation between system parameters and nonlinear dynamic features of multiphysics domains is created, and the principles to ensure high energy oscillation and maximum DC power delivery have been revealed. The established efforts demonstrate a successful synthesis of data-driven and first-principles models to create knowledge in understanding complex dynamic behaviors and facilitate new insights on the implementation of the integrated energy harvesting systems under periodic impulse excitation for enhanced energy conversion efficiency. Moreover, the proposed physics-guided data-driven approach may provide a feasible technique for other real ambient excitations studies in future works.

## 7.3 Impedance analysis for maximizing DC power from nonlinear energy harvesters

Great advancements have been achieved giving attention to nonlinearity in either the structural or electrical sub-systems [9] [42] [52] [137]. Yet, a notable disconnect remains in the knowledge on optimal approaches to integrate nonlinear energy harvesting structures with effective nonlinear rectifying and power management circuits. In Chapters 4 and 5, the impedance model for the nonlinear energy harvester has been formulated for the first time by creating an analogy between the nonlinear mechanical platform and electrical components. The analysis reveals a frequency and amplitude-dependent characteristic of the source impedance due to the nonlinearity introduced to the structures. After establishing the load impedance, a system-level impedance analysis indicates that when the load resistance is much less than the source resistance, the resistance matching plays an essential role in maximizing the delivered DC power. Otherwise, both the reactance and resistance regulate the optimal DC power delivery. Due to the amplitude-dependent characteristic of the source impedance, the conventional impedance matching concept is greatly challenged as the source impedance is a variable. Thus, the optimal strategies for the load resistance around or larger than the source resistance are too complex to identify. In Chapter 5, considering the unpredictability of the kinetic energy [30], a rechargeable battery has been incorporated with high-performance nonlinear interface circuits to scrutinize the principles in maximizing the power delivered from the nonlinear energy harvester. With the source and load impedance models, the impedance analyses conclude that the strategies for maximizing the energy conversion can be determined by solely analyzing the load impedance when charging a battery despite the limitations for the load resistance. These results exemplify the strength of the analytical formulation established here upon impedance principles on examining the strategies on maximizing DC power delivery from the integrated nonlinear energy harvesting systems. The findings may shed light on the principles of the optimal integration of the multiphysics sub-systems to assist in developing sustainable IoT devices.

## 7.4 Establishment of an optimization technique for future balanced nonlinear energy harvesting system design

Energy harvesting systems need to sustainably deliver the required DC power to devices over a long-time horizon. The sole attention on maximizing output voltage or wide frequency range is unable to fulfill the requirements for replacing the consuming batteries [32] [33] [34] [35]. In Chapter 6, an analytical model accounting for the lamination and nonlinearity of the integrated energy harvesting system has been constructed and validated to formulate an optimization approach to exploit nonlinearity in the design of custom-shaped harvesters having excellent DC output power, a broad frequency range of operation, and exceptional mechanical integrity. Both the optimization results and experiment validations indicate that a trapezoidal beam in a monostable configuration is preferred in terms of balancing the frequency bandwidth, DC harvested power, and mechanical integrity. This research provides the first illumination of the coupled influences of nonlinearity and laminated beam shape design on the resulting mechanical integrity and versatile electrical power generation of nonlinear energy harvesting systems. The proposed optimization technique can be utilized to guide the development of integrated energy harvesting systems that yield balanced energy capture performance.

**7.5 Future works**

Motivated by the discoveries and efforts presented in this book, the directions for future work are presented in this section.

**7.5.1 Nonlinear energy harvester integrated with switch-controlled harvesting circuits subjected to real ambient vibrations**

In Chapters 2 and 3, combined harmonic and stochastic excitation and periodic impulse excitation have been taken as representatives of complex vibrations. In future studies, more realistic excitation, such as vibration data recorded from vehicle or machine [172], may be utilized to characterize the nonlinear dynamics and DC power delivery. Moreover, when considering the complex vibration environments, switch-controlled circuits such as SCE circuits and SSHI circuits are rarely considered in studies, whether the vibration energy harvester is operated in a linear or nonlinear regime. The study [173] concluded the outstanding performance of SSHI circuits over the SCE circuit and SEH circuit when interfacing with a linear energy harvester under stochastic excitation through elementary simulations. Yet, the principles to enhance the performance of the integrated systems when the source energy stochasticity and switch-control are incorporated are still lacking. More analytical and numerical efforts should be devoted to investigating the integration of nonlinear energy harvesters and switch-controlled circuits in real ambient vibrations to create knowledge in understanding the intricate dynamics and the influences of the stochasticity of the energy source on the power delivery. The research efforts may assist in the real-world implementation of energy harvesters for improved, robust power delivery.

### 7.5.2 Real-time control for the nonlinear energy harvesting system for maximizing power delivery

Chapter 5 employs impedance principles to investigate power optimization strategies for a nonlinear vibration energy harvester interfaced with a bridge rectifier and a buck-boost converter. The results indicate that the optimal duty cycle for the power management circuit varies with the transition of harmonic excitation conditions, which suggests an open-loop power management circuit cannot sustain high power conversion efficiency. Studies have already proposed an MPPT (maximum power point tracking) control unit for power management circuits to be implemented with a linear energy harvester [174] [175] [176] when subjected to harmonic excitation. Other means, such as reinforcement learning, have also been employed to identify instant resistance utilized for linear energy harvester subjected to aperiodic impulse excitation [177]. Yet, the improvement in energy conversion efficiency that may be achieved by real-time control remains unclear for nonlinear energy harvesters. In future works, attention should be placed to create a closed-loop nonlinear energy harvesting system to shed light on the strategies in maximizing the regulated power through a real-time control approach.

### 7.5.3 Vibration energy harvesting systems with adaptive power management for energy neutral operation

The concept of energy neutral operation (ENO) regards a condition that the energy consumed by the sensor node is less than or equal to the energy harvested by the environment [30]. In the vibration energy harvesting community, continuous efforts have been devoted to enhancing the performance either from the perspective of a mechanical system or electrical system, or both. Yet, the examinations on the neural operation of the vibration energy harvesting system are still lacking. In Chapter 6, we introduced an energy storage component in the energy harvesting system as a first step. In future works, the control of the charging and discharging of the circuit should be implemented to avoid the overcharging or depletion of the battery. In terms of the eventual ENO system design, the statistical characteristics of the vibration excitation conditions and power sensors' working conditions should be examined to estimate the potential energy harvested and consumed. Further, a control algorithm should be exploited to establish the ENO condition [30]. With these efforts, the designed systems are possible to continuously and effectively provide the required power for the load for a long time period.

**Reference**

[1]  H. Sodano, G. Park and D. Inman, "Estimation of electric charge output for piezoelectric energy harvesting," *Strain*, vol. 40(2), pp. 49-58, 2004.

[2]  A. Erturk and D. Inman, "A distributed parameter electromechanical model for cantilevered piezoelectric energy harvesters," *Journal of vibration and acoustics*, vol. 130(4), p. 041002, 2008.

[3]  S. Anton and H. Sodano, "A review of power harvesting using piezoelectric materials (2003–2006)," *Smart materials and Structures*, vol. R1, p. 16(3), 2007.

[4]  M. Lallart, S. Anton and D. Inman, "Frequency self-tuning scheme for broadband vibration energy harvesting," *Journal of Intelligent Material Systems and Structures*, vol. 21(9), pp. 897-906, 2010.

[5]  Z. Xiao, T. Yang, Y. Dong and X. Wang, "Energy harvester array using piezoelectric circular diaphragm for broadband vibration," *Applied Physics Letters*, vol. 104(22), p. 223904, 2014.

[6]  Y. Wu, A. Badel, F. Formosa, W. Liu and A. Agbossou, "Nonlinear vibration energy harvesting device integrating mechanical stoppers used as synchronous mechanical switches," *Journal of Intelligent Material Systems and Structures*, vol. 25(14), pp. 1658-1663, 2014.

[7]  W. Liu, A. Badel, F. Formosa, Y. Wu and A. Agbossou, "Novel piezoelectric bistable oscillator architecture for wideband vibration energy harvesting," *Smart materials and structures*, vol. 22(3), p. 035013, 2013.

[8]  R. Harne and K. Wang, "A review of the recent research on vibration energy harvesting via bistable system," *Smart Materials and Structures*, vol. 22, p. 023001, 2013.

[9]  F. Cottone, H. Vocca and L. Gammaitoni, "Nonlinear energy harvesting," *Physical Review Letters*, vol. 102(8), p. 080601, 2009.

[10]   S. Stanton, C. McGehee and B. Mann, "Nonlinear dynamics for broadband energy harvesting: Investigation of a bistable piezoelectric inertial generator," *Physica D: Nonlinear Phenomena,* vol. 239(10), pp. 640-653, 2010.

[11]   S. Zhou, J. Cao, D. Inman, S. Liu, W. Wang and J. Lin, "Impact-induced high-energy orbits of nonlinear energy harvesters," *Applied Physics Letters,* vol. 106(9), p. 093901, 2015.

[12]   R. Harne, M. Thota and K. Wang, "Concise and high-fidelity predictive criteria for maximizing performance and robustness of bistable energy harvesters," *Applied Physics Letters,* vol. 102(5), p. 053903, 2013.

[13]   R. Harne and K. Wang, "On the fundamental and superharmonic effects in bistable energy harvesting," *Journal of Intelligent Material Systems and Structures,* vol. 25(8), pp. 937-950, 2014.

[14]   Q. Dai and R. Harne, "Investigation of direct current power delivery from nonlinear vibration energy harvesters under combined harmonic and stochastic excitations," *Journal of Intelligent Material Systems and Structures,* p. 1045389X17711788, 2017.

[15]   M. Ferrari, V. Ferrari, M. Guizzetti, B. Ando, S. Baglio and C. Trigona, "Improved energy harvesting from wideband vibrations by nonlinear piezoelectric converters," *Sensors and Actuators A: Physical,* vol. 162, pp. 425-431, 2010.

[16]   Z. Zhou, W. Qin, W. Du, P. Zhu and Q. Liu, "Improving energy harvesting from random excitation by nonlinear flexible bi-stable energy harvester with a variable potential energy function," *Mechanical Systems and Signal Processing,* vol. 115, pp. 162-172, 2019.

[17]   A. Hosseinloo and K. Turitsyn, "Non-resonant energy harvesting via an adaptive bistable potential," *Smart Materials and Structures,* vol. 25(1), p. 015010, 2015.

[18]   W. Wang, J. Cao, C. Bowen, D. Inman and J. Lin, "Performance enhancement of nonlinear asymmetric bistable energy harvesting from harmonic, random and human motion excitations," *Applied Physics Letters,* vol. 112(21), p. 213903, 2018.

[19]   Q. He and M. Daqaq, "Influence of potential function asymmetries on the performance of nonlinear energy harvesters under white noise," *Journal of Sound and Vibration,* vol. 333, pp. 3479-3489, 2014.

[20]   C. Zhang, R. Harne, B. Li and K. Wang, "Statistical quantification of DC power generated by bistable piezoelectric energy harvesters when driven by random excitations," *Journal of Sound and Vibration,* vol. 442, pp. 770-786, 2019.

[21]    Y. Chen, M. Feng and C. Tan, "Modeling of traffic excitation for system identification of bridge structures," *Computer-Aided Civil and Infrastructure Engineering,* vol. 21(1), pp. 57-66, 2006.

[22]    J. Turner and A. Pretlove, "A study of the spectrum of traffic-induced bridge vibration," *Journal of Sound and Vibration,* vol. 122(1), pp. 31-42, 1988.

[23]    K. Ylli, D. Hoffmann, A. Willmann, P. Becker, B. Folkmer and Y. Manoli, "Energy harvesting from human motion: exploiting swing and shock excitations," *Smart Materials and Structures,* vol. 24(2), p. 025029, 2015.

[24]    L. Moro and D. Benasciutti, "Harvested power and sensitivity analysis of vibrating shoe-mounted piezoelectric cantilevers," *Smart Materials and Structures,* vol. 19, no. 11, p. 115011, 2010.

[25]    A. Doria, C. Medè, D. Desideri, A. Maschio, L. Codecasa and F. Moro, "On the performance of piezoelectric harvesters loaded by finite width impulses," *Mechanical Systems and Signal Processing,* vol. 100, pp. 28-42, 2018.

[26]    J. Liang and W. Liao, "Impedance modeling and analysis for piezoelectric energy harvesting systems," *IEEE/ASME Transactions on Mechatronics ,* vol. 17(6), pp. 1145-1157, 2012.

[27]    Y. Liao and J. Liang, "Unified modeling analysis and comparison of piezoelectric vibration energy harvesters," *Mechanical Systems and Signal Processing,* vol. 123, pp. 403-425, May 2019.

[28]    Y. Liao and J. Liang, "Maximum power, optimal load, and impedance analysis of piezoelectric vibration energy harvesters," *Smart Materials and Structures,* vol. 27(7), p. 075053, 2018.

[29]    F. Shaikh and S. Zeadally, "Energy harvesting in wireless sensor networks: A comprehensive review," *Renewable and Sustainable Energy Reviews,* vol. 55, pp. 1041-1054, 2016.

[30]    S. Shresthamali, M. Kondo and H. Nakamura, "Adaptive power management in solar energy harvesting sensor node using reinforcement learning," *ACM Transactions on Embedded Computing Systems (TECS),* vol. 16(3s), pp. 1-21, 2017.

[31]    K. Savarimuthu, R. Sankararajan and S. Murugesan, "Design and implementation of piezoelectric energy harvesting circuit," *Circuit World,* 2017.

[32]    J. Dietl and E. Garcia, "Beam shape optimization for power harvesting," *Journal of Intelligent Material Systems and Structures,* vol. 21(6), pp. 633-646, 2010.

[33]    J. Cho, M. Anderson, R. Richards, D. Bahr and C. Richards, "Optimization of electromechanical coupling for a thin-film PZT membrane: I. modeling," *Journal of Micromechanics and Microengineering,* vol. 15(10), p. 1797, 2005.

[34]    Q. Wang and N. Wu, "Optimal design of a piezoelectric coupled beam for power harvesting," *Smart Materials and Structures,* vol. 21(8), p. 085013, 2012.

[35]    C. Lü, Y. Zhang, H. Zhang, Z. Zhang, M. Shen and Y. Chen, "Generalized optimization method for energy conversion and storage efficiency of nanoscale flexible piezoelectric energy harvesters," *Energy Conversion and Management,* vol. 182, pp. 34-40, 2019.

[36]    M. Jia, A. Komeily, Y. Wang and R. Srinivasan, "Adopting Internet of Things for the development of smart buildings: A review of enabling technologies and applications," *Automation in Construction,* vol. 101, pp. 111-126, 2019.

[37]    A. Darwish, A. Hassanien, M. Elhoseny, A. Sangaiah and K. Muhammad, "The impact of the hybrid platform of internet of things and cloud computing on healthcare systems: Opportunities, challenges, and open problems," *Journal of Ambient Intelligence and Humanized Computing,* pp. 1-16, 2017.

[38]    M. Díaz, C. Martín and B. Rubio, "State-of-the-art, challenges, and open issues in the integration of Internet of things and cloud computing," *Journal of Network and Computer applications,* vol. 67, pp. 99-117, 2016.

[39]    L. Reindl, "Power supply for wireless sensor systems," 2018. [Online]. Available: http://www.sensornets.org/Documents/Previous_Invited_Speakers/2018/SENSORNETS2 018_Reindl.pdf.

[40]    A. Hande, R. Bridgelall and B. Zoghi, "Vibration energy harvesting for disaster asset monitoring using active RFID tags," *Proceedings of the IEEE,* vol. 98(9), pp. 1620-1628, 2010.

[41]    A. Erturk and D. Inman, "Broadband piezoelectric power generation on high-energy orbits of the bistable duffing oscillator with electromechanical coupling," *Journal of Sound and Vibration,* vol. 330, pp. 2339-2353, 2011.

[42]    A. Erturk, J. Hoffmann and D. Inman, "A piezomagnetoelastic structure for broadband vibration energy harvesting," *Applied Physics Letters,* vol. 94, p. 254102, 2009.

[43] T. Huguet, M. Lallart and A. Badel, "Orbit jump in bistable energy harvesters through buckling level modification," *Mechanical Systems and Signal Processing,* vol. 128, pp. 202-215, 2019.

[44] J. Wang and W. Liao, "Attaining the high-energy orbit of nonlinear energy harvesters by load perturbation," *Energy Conversion and Management,* vol. 192, pp. 30-36, 2019.

[45] R. Hosseini and M. Hamedi, "Improvements in energy harvesting capabilities by using different shapes of piezoelectric bimorphs," *Journal of Micromechanics and Microengineering,* vol. 25(12), p. 125008, 2015.

[46] B. Montazer and U. Sarma, "Design and Optimization of Quadrilateral Shaped PVDF Cantilever for Efficient Conversion of Energy from Ambient Vibration," *IEEE Sensors Journal,* vol. 18(10), pp. 3977-3988, 2018.

[47] W. Cai and R. Harne, " Vibration energy harvesters with optimized geometry, design, and nonlinearity for robust direct current power delivery," *Smart Materials and Structures,* 2019.

[48] W. Cai and R. L. Harne, "Optimized piezoelectric energy harvesters for performance robust operation in periodic vibration environments," in *Active and Passive Smart Structures and Integrated Systems XII* , Denver,Colorado, 2019.

[49] W. Liu, C. Zhao, A. Badel, F. Formosa, Q. Zhu and G. Hu, "Compact self-powered synchronous energy extraction circuit design with enhanced performance," *Smart Materials and Structures,* vol. 27(4), p. 047001, 2018.

[50] W. Cai and R. Harne, "Electrical power management and optimization with nonlinear energy harvesting structures," *Journal of Intelligent Material Systems and Structures,* vol. 30(2), pp. 213-227, 2019.

[51] E. Lefeuvre, A. Badel, C. Richard and D. Guyomar, "Piezoelectric energy harvesting device optimization by synchronous electric charge extraction," *Journal of Intelligent Material Systems and Structures,* vol. 16(10), pp. 865-876, 2005.

[52] E. Lefeuvre, A. Badel, A. Brenes, S. Seok and C. Yoo, "Power and frequency bandwidth improvement of piezoelectric energy harvesting devices using phase-shifted synchronous electric charge extraction interface circuit," *Journal of Intelligent Material Systems and Structures,* vol. 28(20), pp. 2988-2995, 2017.

[53]   N. Anh and N. Hieu, "The Duffing oscillator under combined periodic and random excitations," *Probabilistic Engineering Mechanics*, vol. 30, pp. 27-36, 2012.

[54]   R. Iyengar, " A nonlinear system under combined periodic and random excitation," *Journal of statistical physics*, Vols. 44(5-6), pp. 907-920, 1986.

[55]   Z. Huang, W. Zhu and Y. Suzuki, "Stochastic averaging of strongly non-linear oscillators under combined harmonic and white-noise excitations," *Journal of Sound and Vibration*, vol. 238(2), pp. 233-256, 2000.

[56]   Y. Wu and W. Zhu, " Stochastic averaging of strongly nonlinear oscillators under combined harmonic and wide-band noise excitations," *Journal of Vibration and Acoustics*, vol. 130(5), p. 051004, 2008.

[57]   A. Bulsara, K. Lindenberg and K. Shuler, "Spectral analysis of a nonlinear oscillator driven by random and periodic forces. I. Linearized theory," *Journal of Statistical Physics*, vol. 27(4), pp. 787-808, 1982.

[58]   H. Kim, W. Tai, J. Parker and L. Zuo, "Self-tuning stochastic resonance energy harvesting for rotating systems under modulated noise and its application to smart tires," *Mechanical Systems and Signal Processing*, vol. 122, pp. 769-785, 2019.

[59]   S. Zhou and L. Zuo, " Nonlinear dynamic analysis of asymmetric tristable energy harvesters for enhanced energy harvesting," *Communications in Nonlinear Science and Numerical Simulation*, vol. 61, pp. 271-284, 2018.

[60]   S. Fang, X. Fu and W. Liao, "Asymmetric plucking bistable energy harvester: Modeling and experimental validation," *Journal of Sound and Vibration*, vol. 459, p. 114852, 2019.

[61]   S. Ali, S. Adhikari, M. Friswell and S. Narayanan, "The analysis of piezomagnetoelastic energy harvesters under broadband random excitations," *Journal of Applied Physics*, vol. 109(7), p. 074904, 2011.

[62]   Y. Shu and I. Lien, "Analysis of power output for piezoelectric energy harvesting systems," *Smart materials and structures*, vol. 15(6), p. 1499, 2006.

[63]   W. Jiang and L. Chen, "Stochastic averaging based on generalized harmonic functions for energy harvesting systems," *Journal of Sound and Vibration*, vol. 377, pp. 264-283, 2016.

[64]   Z. Xu and Y. Cheung, "Averaging method using generalized harmonic functions for strongly non-linear oscillators," *Journal of Sound and Vibration,* vol. 174(4), pp. 563-576, 1994.

[65]   H. Makarem, H. N. Pishkenari and G. R. Vossoughi, "A modified Gaussian moment closure method for nonlinear stochastic differential equations," *Nonlinear Dynamics,* vol. 89(4), pp. 2609-2620, 2017.

[66]   H. Zhu and S. Guo, "Periodic response of a Duffing oscillator under combined harmonic and random excitations," *Journal of Vibration and Acoustics,* vol. 137(4), 2015.

[67]   A. Erturk and D. Inman, Piezoelectric energy harvesting, Chichester: John Wiley & Sons, 2011.

[68]   Q. Dai and R. Harne, "Charging power optimization for nonlinear vibration energy harvesting systems subjected to arbitrary, persistent base excitations," *Smart Materials and Structures,* vol. 27(1), p. 015011, 2017.

[69]   J. A. Hansen and C. Penland, "Efficient approximate techniques for integrating stochastic differential equations," *Monthly weather review,* vol. 134(10), pp. 3006-3014, 2006.

[70]   M. Dykman and K. Lindenberg, "Fluctuations in nonlinear systems driven by colored noise," in *Contemporary Problems in Statistical Physics*, G. Weiss, Ed., Philadelphia, SIAM, 1994.

[71]   S. Zeadally, F. Shaikh, A. Talpur and Q. Sheng, "Design architectures for energy harvesting in the Internet of Things," *Renewable and Sustainable Energy Reviews,* vol. 128, p. 109901, 2020.

[72]   Z. Xiao, T. Yang and Y. Dong, "Energy harvester array using piezoelectric circular diaphragm for broadband vibration," *Applied Physics Letters,* vol. 104, no. 22, p. 223904, 2014.

[73]   D. Mallick, A. Amann and S. Roy, "Surfing the high energy output branch of nonlinear energy harvesters," *Physical review letters,* vol. 117, no. 19, p. 197701, 2016.

[74]   J. Wang, B. Zhao, W. Liao and J. Liang, "New insight into piezoelectric energy harvesting with mechanical and electrical nonlinearities," *Smart Materials and Structures,* vol. 29, no. 4, p. 04LT01, 2020.

[75] S. Leadenham and A. Erturk, "Mechanically and electrically nonlinear non-ideal piezoelectric energy harvesting framework with experimental validations," *Nonlinear Dynamics,* vol. 99, no. 1, pp. 625-641, 2020.

[76] T. Huguet, M. Lallart and A. Badel, "Bistable vibration energy harvester and SECE circuit: exploring their mutual influence," *Nonlinear Dynamics,* vol. 97, no. 1, pp. 485-501, 2019.

[77] X. Rui, Y. Zhang, Z. Zeng, G. Yue, X. Huang and J. Li, "Design and analysis of a broadband three-beam impact piezoelectric energy harvester for low-frequency rotational motion," *Mechanical Systems and Signal Processing,* vol. 149, p. 107307, 2021.

[78] R. Harne, C. Zhang, B. Li and K. Wang, "An analytical approach for predicting the energy capture and conversion by impulsively-excited bistable vibration energy harvesters," *Journal of Sound and Vibration,* vol. 373, pp. 205-222, 2016.

[79] C. Zhang, R. Harne, B. Li and K. Wang, "Reconstructing the transient, dissipative dynamics of a bistable Duffing oscillator with an enhanced averaging method and Jacobian elliptic functions," *International Journal of Non-Linear Mechanics,* vol. 79, pp. 26-37, 2016.

[80] K. Remick, H. Joo, D. McFarland, T. Sapsis, L. Bergman, D. Quinn and A. Vakakis, "Sustained high-frequency energy harvesting through a strongly nonlinear electromechanical system under single and repeated impulsive excitations," *Journal of Sound and Vibration,* vol. 333, no. 14, pp. 3214-3235, 2014.

[81] X. Fu and W. Liao, "Nondimensional model and parametric studies of impact piezoelectric energy harvesting with dissipation," *Journal of Sound and Vibration,* vol. 429, pp. 78-95, 2018.

[82] D. Quinn, A. Triplett, A. Vakakis and L. Bergman, "Energy harvesting from impulsive loads using intentional essential nonlinearities," *Journal of Vibration and Acoustics,* vol. 133, no. 1, 2011.

[83] S. Chiacchiari, F. Romeo, D. McFarland, L. Bergman and A. Vakakis, "Vibration energy harvesting from impulsive excitations via a bistable nonlinear attachment," *International Journal of Non-Linear Mechanics,* vol. 94, pp. 84-97, 2017.

[84] A. Karpatne, W. Watkins, J. Read and V. Kumar, "Physics-guided neural networks (pgnn): An application in lake temperature modeling," *arXiv preprint arXiv,* vol. 1710, p. 11431, 2017.

[85] Y. Yu, H. Yao and Y. Liu, "Structural dynamics simulation using a novel physics-guided machine learning method," *Engineering Applications of Artificial Intelligence,* vol. 96, p. 103947, 2020.

[86] N. Wagner and J. Rondinelli, "Theory-guided machine learning in materials science," *Frontiers in Materials,* vol. 3, p. 28, 2016.

[87] S. Brunton, J. Proctor and J. Kutz, "Discovering governing equations from data by sparse identification of nonlinear dynamical systems," *Proceedings of the National Academy of Sciences,* vol. 113, no. 15, pp. 3932-3937, 2016.

[88] Z. Zhang and C. Sun, "Structural damage identification via physics-guided machine learning: a methodology integrating pattern recognition with finite element model updating," *Structural Health Monitoring,* p. 1475921720927488, 2020.

[89] J. Ling, R. Jones and J. Templeton, "Machine learning strategies for systems with invariance properties," *Journal of Computational Physics,* vol. 318, pp. 22-35, 2016.

[90] Q. Dai, I. Park and R. Harne, "Impulsive energy conversion with magnetically coupled nonlinear energy harvesting systems," *Journal of Intelligent Material Systems and Structures,* vol. 29, no. 11, pp. 2374-2391, 2018.

[91] J. Wang, D. Sun, S. Liu and X. Zhang, "Damping characteristics of viscoelastic damping structure under coupled condition," *Mathematical and Computational Applications,* vol. 22, no. 1, p. 27, 2017.

[92] C. Vasques, R. Moreira and J. Rodrigues, "Viscoelastic Damping Technologies-Part I: Modeling and Finite Element Implementation," *Journal of Advanced Research in Mechanical Engineering,* vol. 1, no. 2, 2010.

[93] J. Willard, X. Jia and S. Xu, "Integrating physics-based modeling with machine learning: A survey," *arXiv preprint arXiv:2003,* p. 04919, 2020.

[94] G. Chen, Y. Zuo and J. Sun, "Support-vector-machine-based reduced-order model for limit cycle oscillation prediction of nonlinear aeroelastic system," *Mathematical problems in engineering,* p. 2012, 2012.

[95] J. Mahanta, July 2017. [Online]. Available: https://towardsdatascience.com/introduction-to-neural-networks-advantages-and-applications-96851bd1a207.

[96] E. Alpaydin, Introduction to Machine Learning, 3rd ed., The MIT Press, 2014.

[97]  M. Mishra, June 2018. [Online]. Available: https://towardsdatascience.com/classification-an-important-concept-in-machine-learning-af6ff4cb2cfd.

[98]  M. Cao, N. Alkayem, L. Pan and D. Novak, "Advanced methods in neural networks-based sensitivity analysis with their applications in civil engineering," *Artificial neural networks: models and applications, Rijeka, Croatia, IntechOpen,* pp. 335-353, 2016.

[99]  S. Beeby, M. Tudor and N. White, "Energy harvesting vibration sources for microsystems applications," *Measurement Science and Technology,* vol. 17, pp. R175-R195, 2006.

[100]  P. Mitcheson, E. Yeatman, G. Rao, A. Holmes and T. Green, "Energy harvesting from human and machine motion for wireless electronic devices," *Proceedings of the IEEE,* vol. 96, pp. 1457-1486, 2008.

[101]  M. Shafer, R. MacCurdy, J. Shipley, D. Winkler, C. Guglielmo and E. Garcia, "The case for energy harvesting on wildlife in flight," *Smart Materials and Structures,* vol. 24, p. 025031, 2015.

[102]  H. Jawad, R. Nordin, S. Gharghan, A. Jawad and M. Ismail, "Energy-Efficient wireless sensor networks for precision agriculture: A review," *Sensors,* vol. 17, p. 1781, 2017.

[103]  A. Chandrakasan, R. Amirtharajah, J. Goodman and W. Rabiner, "Trends in low power digital signal processing," *Proceedings of the 1998 IEEE International Symposium on Circuits and Systems,* vol. 4, pp. 604-607, 1998.

[104]  W. Davis, N. Zhang, K. Camera, F. Chen, D. Markovic, N. Chan, B. Nikolic and R. Brodersen, "A design environment for high-throughput low power dedicated siganl processing systems," *IEEE Journal of Solid-State Circuits,* vol. 37, pp. 420-431, 2002.

[105]  S. Roundy and Y. Zhang, "Toward self-tuning adaptive vibration based micro-generators," *Smart Structures, Devices, and Systems II. International Society for Optics and Photonics.,* vol. 5649, pp. 373-385, 2005.

[106]  G. Sebald, H. Kuwano, D. Guyomar and B. Ducharne, "Experimental duffing oscillator for broadband piezoelectric energy harvesting," *Smart Materials and Structures,* vol. 20, p. 102001, 2011.

[107]  L. Gu and C. Livermore, "Impact-driven, frequency up-converting coupled vibration energy harvesting device for low frequency operation," *Smart Materials and Structures,* vol. 20, p. 045004, 2011.

[108]  G. Scarselli, F. Nicassio, F. Pinto, F. Ciampa, O. Lervolino and M. Meo, "A novel bistable energy harvesting concept," *Smart Materials and Structures,* vol. 25, p. 055001, 2016.

[109]  G. Szarka, B. Stark and S. Burrow, "Review of power conditioning for kinetic energy harvesting systems," *IEEE Transactions on Power Electronics,* vol. 27(2), pp. 803-815, 2012.

[110]  T. Le, J. Han, A. von Jouanne, K. Mayaram and T. Fiez, "Piezoelectric micro-power generation interface circuits," *IEEE Journal of Solid-State Circuits,* vol. 41(6), pp. 1411-1420, 2006.

[111]  Y. Lam, W. Ki and C. Tsui, "Integrated low-loss CMOS active rectifier for wirelessly powered devices," *IEEE Transactions on Circuits and Systems II: Express Briefs,* vol. 53(12), pp. 1378-1382, 2006.

[112]  D. Guyomar, A. Badel and E. Lefeuvre, "Toward energy harvesting using active materials and conversion improvement by nonlinear processing," *IEEE transactions on ultrasonics, ferroelectrics, and frequency control,* vol. 52, pp. 584-595, 2005.

[113]  E. Lefeuvre, A. Badel, C. Richard, L. Petit and D. Guyomar, "A comparison between several vibration-powered piezoelectric generators for standalone systems," *Sensors and Actuators A: Physical,* vol. 126(2), pp. 405-416, 2006.

[114]  K. Makihara, J. Onoda and T. Miyakawa, "Low energy dissipation electric circuit for energy harvesting," *Smart Materials and Structures,* vol. 15, pp. 1493-1498, 2006.

[115]  L. Garbuio, M. Lallart, D. Guyomar, C. Richard and D. Audigier, "Mechanical energy harvester with ultralow threshold rectification based on SSHI nonlinear technique," *IEEE Transactions on Industrial Electronics,* vol. 56, pp. 1048-1056, 2009.

[116]  G. Ottman, H. Hofmann, A. Bhatt and G. Lesieutre, "Adaptive piezoelectric energy harvesting circuit for wireless remote power supply," *IEEE Transactions on Power Electronics,* vol. 17, pp. 669-676, 2002.

[117]  F. Moon and P. Holmes, "A magnetoelastic strange attractor," *Journal of Sound and Vibration,* vol. 65(2), pp. 275-296, 1979.

[118]  T. Hikihara and T. Kawagoshi, "An experimental study on stablilization of unstable periodic motion in magneto-elastic chaos," *Physics Letters A,* vol. 211(1), pp. 29-36, 1996.

[119] B. Feeny, C. Yuan and J. Cusumano, "Parametric identification of an experimental magneto-elastic oscillator," *Journal of Sound and Vibration,* vol. 247, pp. 785-806, 2001.

[120] N. Kong, D. Ha, A. Erturk and D. Inman, "Resistive impedance matching circuit for piezoelectric energy harvesting," *Journal of Intelligent Material Systems and Structures,* vol. 21, pp. 1293-1302, 2010.

[121] Y. Yang and L. Tang, "Equivalent circuit modeling of piezoelectric energy harvesters," *Journal of Intelligent Material Systems and Structures,* vol. 20, pp. 2223-2235, 2009.

[122] R. Harne and K. Wang, Harnessing Bistable Structural Dynamics: For Vibration Control, Energy Harvesting and Sensing, Chichester: John Wiley & Sons, 2017.

[123] E. Lefeuvre, D. Audigier, C. Richard and D. Guyomar, "Buck-boost converter for sensorless power optimization of piezoelectric energy harvester," *IEEE Transactions on Power Electronics,* vol. 22(5), pp. 2018-2025, 2007.

[124] E. Rogers, "Understanding buck-boost power stages in switch mode power supplies," 1999.

[125] M. Panyam, R. Masana and M. Daqaq, "On approximating the effective bandwidth of bi-stable energy harvesters," *International Journal of Non-Linear Mechanics,* vol. 67, pp. 153-163, 2014.

[126] I. Kovacic, M. Brennan and B. Lineton, "On the resonance response of an asymmetric duffing oscillator," *International Journal of Non-Linear Mechanics,* vol. 43, pp. 858-867, 2008.

[127] B. Goodpaster and R. Harne, "Analytical Modeling and Impedance Characterization of the Nonlinear Dynamics of Thermomechanically Coupled Structures," *Journal of Applied Mechanics,* vol. 85, p. 081010, 2018.

[128] D. Maayan, January 2020. [Online]. Available: https://securitytoday.com/Articles/2020/01/13/The-IoT-Rundown-for-2020.aspx?Page=2.

[129] A. Aman, N. Shaari and R. Ibrahim, "Internet of things energy system: Smart applications, technology advancement, and open issues," *International Journal of Energy Research,* 2021.

[130] X. Wang and B. Mann, "Attractor Selection in Nonlinear Energy Harvesting Using Deep Reinforcement Learning," *arXiv preprint arXiv,* p. 2010.01255, 2020.

[131] H. Pan, L. Qi, Z. Zhang and J. Yan, "Kinetic energy harvesting technologies for applications in land transportation: A comprehensive review," *Applied Energy*, vol. 286, p. 116518, 2021.

[132] D. Oletic and V. Bilas, "System-level power consumption analysis of the wearable asthmatic wheeze quantification," *Journal of Sensors*, p. 2018, 2018.

[133] W. Cai and R. Harne, "Characterization of challenges in asymmetric nonlinear vibration energy harvesters subjected to realistic excitation," *Journal of Sound and Vibration*, p. 115460, 2020.

[134] Z. Zhang, H. Xiang and L. Tang, "Modeling, analysis and comparison of four charging interface circuits for piezoelectric energy harvesting," *Mechanical Systems and Signal Processing*, vol. 152, p. 107476, 2021.

[135] H. Sodano, G. Park, D. Leo and D. Inman, "Use of piezoelectric energy harvesting devices for charging batteries," in *SPIE 10th Annual International Symposium on Smart Structures and Materials*, San Diego, 2003.

[136] H. Hu, H. Xue and Y. Hu, "A spiral-shaped harvester with an improved harvesting element and an adaptive storage circuit," *IEEE transactions on ultrasonics, ferroelectrics*, vol. 54(6), pp. 1177-1187, 2007.

[137] Y. Shu, I. Lien and W. Wu, "An improved analysis of the SSHI interface in piezoelectric energy harvesting," *Smart Materials and Structures*, vol. 16, pp. 2253-2264, 2007.

[138] E. Kim, "Thevenin's and Norton's Equivalent Circuit Tutorial," [Online]. Available: https://alan.ece.gatech.edu/ECE3040/Lectures/CircuitReview.pdf.

[139] C. Chen, K. Zhao and J. Liang, "Impedance analysis of piezoelectric energy harvesting system using synchronized charge extraction interface circuit," in *Active and Passive Smart Structures and Integrated Systems 2017* , Portland, 2017.

[140] J. Liang and W. Liao, "Piezoelectric energy harvesting and dissipation on structural damping," *Journal of Intelligent Material Systems and Structures*, vol. 20(5), pp. 515-527, 2009.

[141] M. Magno, L. Spadaro, J. Singh and L. Benini, "Kinetic energy harvesting: toward autonomous wearable sensing for internet of things," in *2016 International Symposium on Power Electronics, Electrical Drives, Automation and Motion*, Anacapri, 2016.

[142] D. Evans, "The internet of things: how the next evolution of the internet is changing everything," *CISCO White Paper,* vol. 1, pp. 1-11, 2011.

[143] H. Jayakumar, K. Lee, W. Lee, A. Raha, Y. Kim and V. Raghunathan, "Powering the internet of things," in *Proceedings of the 2014 International Symposium on Low Power Electronics and Design,* La Jolla, 2014.

[144] M. Gorlatova, J. Sarik, G. Grebla, M. Cong, I. Kymissis and G. Zussman, "Movers and shakers: kinetic energy harvesting for the internet of things," *ACM SIGMETRICS Performance Evaluation Review,* vol. 42(1), pp. 407-419, 2014.

[145] L. Qin, J. Jia, M. Choi and K. Uchino, " Improvement of electromechanical coupling coefficient in shear-mode of piezoelectric ceramics," *Ceramics International,* vol. 45(2), pp. 1496-1502, 2019.

[146] J. Baker, S. Roundy and P. Wright, "Alternative geometries for increasing power density in vibration energy scavenging for wireless sensor networks," in *3rd International Energy Conversion Engineering Conference,* San Francisco, 2005.

[147] S. Roundy, E. Leland, J. Baker, E. Carleton, E. Reilly, E. Lai, B. Otis, J. Rabaey, V. Sundararajan and P. Wright, "Improving power output for vibration-based energy scavengers," *IEEE Pervasive Computing,* vol. 1(1), pp. 28-36, 2005.

[148] S. Tabatabaei, S. Behbahani and P. Rajaeipour, "Multi-objective shape design optimization of piezoelectric energy harvester using artificial immune system," *Microsystem Technologies,* vol. 22(10), pp. 2435-2446, 2016.

[149] S. Raju, M. Umapathy and G. Uma, "High-output piezoelectric energy harvester using tapered beam with cavity," *Journal of Intelligent Material Systems and Structures,* vol. 29(5), pp. 800-815, 2018.

[150] J. Xu, Y. Liu, W. Shao and Z. Feng, "Optimization of a right-angle piezoelectric cantilever using auxiliary beams with different stiffness levels for vibration energy harvesting," *Smart Materials and Structures,* vol. 21(6), p. 065017, 2012.

[151] H. Yoon, G. Washington and A. Danak, "Modeling, optimization, and design of efficient initially curved piezoceramic unimorphs for energy harvesting applications," *Journal of Intelligent Material Systems and Structures,* vol. 16(10), pp. 877-888, 2005.

[152] N. Chandrasekharan and L. Thompson, " Increased power to weight ratio of piezoelectric energy harvesters through integration of cellular honeycomb structures," *Smart Materials and Structures,* vol. 25(4), p. 045019, 2016.

[153] M. Nguyen, Y. Yoon, O. Kwon and P. Kim, "Lowering the potential barrier of a bistable energy harvester with mechanically rectified motion of an auxiliary magnet oscillator," *Applied Physics Letters,* vol. 111(25), p. 253905, 2017.

[154] W. Yang and S. Towfighian, "Low frequency energy harvesting with a variable potential function under random vibration," *Smart Materials and Structures,* vol. 27(11), p. 114004, 2018.

[155] M. Panyam and M. Daqaq, "Characterizing the effective bandwidth of tri-stable energy harvesters," *Journal of Sound and Vibration,* vol. 386, pp. 336-358, 2017.

[156] S. Lai, C. Wang and L. Zhang, " A nonlinear multi-stable piezomagnetoelastic harvester array for low-intensity, low-frequency, and broadband vibrations," *Mechanical Systems and Signal Processing,* vol. 122, pp. 87-102, 2019.

[157] P. Avvari, Y. Yang and C. Soh, " Long-term fatigue behavior of a cantilever piezoelectric energy harvester," *Journal of Intelligent Material Systems and Structures,* vol. 28(9), pp. 1188-1210, 2017.

[158] D. Upadrashta, Y. Yang and L. Tang, "Material strength consideration in the design optimization of nonlinear energy harvester," *Journal of Intelligent Material Systems and Structures,* vol. 26(15), pp. 1980-1994, 2015.

[159] A. Pandey and A. Arockiarajan, "An experimental and theoretical fatigue study on macro fiber composite (MFC) under thermo-mechanical loadings," *European Journal of Mechanics-A/Solids,* vol. 66, pp. 26-44, 2017.

[160] S. Gundimeda, S. Kunc, J. Gallagher and R. Fragoudakis, "Simulation of a Composite Piezoelectric and Glass Fiber Reinforced Polymer Beam for Adaptive Stiffness Applications," in *ASME 2018 Conference on Smart Materials, Adaptive Structures and Intelligent Systems. American Society of Mechanical Engineers*, San Antonio, 2018.

[161] X. Li, D. Upadrashta, K. Yu and Y. Yang, "Sandwich piezoelectric energy harvester: Analytical modeling and experimental validation," *Energy Conversion and Management,* vol. 176, pp. 69-85, 2018.

[162] R. Harne and K. Wang, "Axial suspension compliance and compression for enhancing performance of a nonlinear vibration energy harvesting beam system," *Journal of Vibration and Acoustics,* vol. 138(1), p. 011004, 2016.

[163] F. Goldschmidtboeing and P. Woias, "Characterization of different beam shapes for piezoelectric energy harvesting," *Journal of Micromechanics and Microengineering,* vol. 18(10), p. 104013, 2008.

[164] N. Hagood, W. Chung and A. Von Flotow, "Modelling of piezoelectric actuator dynamics for active structural control," *Journal of Intelligent Material Systems and Structures,* vol. 1(3), pp. 327-354, 1990.

[165] C. Lan, L. Tang and R. Harne, "Comparative methods to assess harmonic response of nonlinear piezoelectric energy harvesters interfaced with AC and DC circuits," *Journal of Sound and Vibration,* vol. 421, pp. 61-78, 2018.

[166] J. Jaworski and E. Dowell, "Free vibration of a cantilevered beam with multiple steps: Comparison of several theoretical methods with experiment," *Journal of Sound and Vibration,* Vols. 312(4-5), pp. 713-725, 2008.

[167] P. Spanos, "Stochastic linearization in structural dynamics," *Applied Mechanics Reviews,* vol. 34(1), pp. 1-8, 1981.

[168] R. Harne and B. Goodpaster, "Impedance measures in analysis and characterization of multistable structures subjected to harmonic excitation," *Mechanical Systems and Signal Processing,* vol. 98, pp. 78-90, 2018.

[169] T. Atalik and S. Utku, "Stochastic linearization of multi-degree-of-freedom non-linear systems," *Earthquake Engineering & Structural Dynamics 4,* vol. 4, pp. 411-420, 1976.

[170] R. Meier, N. Kelly, O. Almog and P. Chiang, "A piezoelectric energy-harvesting shoe system for podiatric sensing," in *Engineering in Medicine and Biology Society (EMBC), 2014 36th Annual International Conference of the IEEE,* Chicago, 2014.

[171] R. Haupt and S. Haupt, Practical Genetic Algorithms, New York: Wiley, 1998.

[172] R. Rantz and S. Roundy, "Characterization of real-world vibration sources and application to nonlinear vibration energy harvesters," *Energy Harvesting and Systems,* vol. 4(2), pp. 67-76, 2017.

[173] S. Pang, W. Li and J. Kan, "Simulation Analysis of Interface Circuits for Piezoelectric Energy Harvesting with Damped Sinusoidal Signals and Random Signals," *Telkomnika*, vol. 13(3), p. 767, 2015.

[174] N. Kong and D. Ha, "Low-power design of a self-powered piezoelectric energy harvesting system with maximum power point tracking," *IEEE Transactions on power electronics*, vol. 27(5), pp. 2298-2308, 2011.

[175] S. Li, A. Roy and B. Calhoun, "A Piezoelectric Energy-Harvesting System with Parallel-SSHI Rectifier and Integrated MPPT Achieving 417% Energy-Extraction Improvement and 97% Tracking Efficiency," in *2019 Symposium on VLSI Circuits*, Kyoto, 2019.

[176] G. Shi, Y. Xia, H. Xia, X. Wang, L. Qian, Z. Chen and Q. Li, "An efficient power management circuit based on quasi maximum power point tracking with bidirectional intermittent adjustment for vibration energy harvesting," *IEEE Transactions on Power Electronics*, vol. 34, no. 10, pp. 9671-9685, October 2019.

[177] J. Ward and S. Behrens, "Adaptive learning algorithms for vibration energy harvesting," *Smart Materials and Structures*, vol. 17(3), p. 035025, 2008.